Table of Contents

List of Tables

List of Figures

Executive Summary

This report summarizes the offshore wind resource potential, based on map estimates, for the contiguous United States and Hawaii, as of May 2009. The development of this assessment has evolved over multiple stages as new regional meso-scale assessments became available, new validation data were obtained, and better modeling capabilities were implemented. It is expected that further updates to the current assessment will be made in future reports.

Offshore wind energy development promises to be a significant domestic renewable energy source, especially for coastal energy loads with limited access to interstate grid transmission. The definition of the magnitude and distribution of this resource required the development of a standard and flexible database. Developed using Geographic Information System (GIS) techniques, the database includes offshore wind resource characteristics such as wind speed, water depth, and distance from shore. It combines the resource characteristics with state administrative areas and quantifies the resource for several scenarios. In the future, the database may be expanded to include other important characteristics such as wave power density, extreme wind and wave, ocean currents, and a number of other parameters important to the design of offshore wind turbines.

The primary method used to present the offshore wind resource data are maps that categorize the resource by annual average wind speed at 90 meters (m) above the surface. The resource maps extend from the shoreline out to 50 nautical miles (nm) offshore. Exceptions to the 50 nm mapped distance are the Great Lakes that were mapped in their entirety for the offshore resource and Massachusetts, where the computed resource did not extend 50 nm from the edge of the Nantucket Island and Martha's Vineyard in southeastern Massachusetts. The offshore maps for some states do not extend 50 nm because of state and administrative boundaries.

There were several sources for the offshore resource maps. The oldest offshore wind resource maps were generated from data mapped as part of onshore state wind resource mapping projects. Recently, some regions have had their resource maps updated using the latest computer mapping techniques optimized for offshore modeling. For this evaluation, updated maps were available for the offshore areas of Georgia, Texas/Louisiana, northern New England (Massachusetts, New Hampshire, and Maine), and states that border the Great Lakes. The preliminary numerical modeling was performed by AWS Truepower (AWST), of Albany, New York, under subcontract to the National Renewable Energy Laboratory (NREL), using their proprietary MesoMap system. The preliminary model estimates were validated by NREL using data from a variety of sources including ocean buoys, marine automated stations, Coast Guard stations and lighthouses, and satellite-derived 10-m wind speeds over the ocean estimated from the "state of the sea" as measured by microwave imaging. Final modifications to the preliminary estimates were agreed to after consultations between NREL and AWST. AWST adjusted the model output to reflect the modifications and sent the final grids to NREL, where the data were converted into wind resource maps. The calculation of the offshore wind resource estimates for states without updated maps depended on the availability of older offshore wind maps that were completed as part of onshore wind mapping projects. Where older preliminary and final offshore wind maps were available, the wind speed data from those maps were interpolated to 90 m height and, if necessary, extrapolated to the 50 nm lines. The offshore areas treated in this manner included the Atlantic coast, from Rhode Island to South Carolina, the Pacific coast, from California to Washington, and Hawaii. The states of Florida, Alabama, and Mississippi did not have any preliminary or final wind maps available. The offshore wind resource for these states is not included in this report.

Table 1 shows the offshore wind resource by available square kilometers (km^2) of water and potential installed capacity in gigawatts (GW) for annual average wind speeds greater than 7.0 meters/second (m/s) at 90 m above the surface. A uniform factor of 5 megawatts/km^2 was applied to calculate the potential installed capacity. The resource is presented for individual states and the country as a whole. These resource estimates have not been reduced by any environmental or water-use considerations. Detailed information by database element for each state is presented in Appendix B. The data presented in this report represents the first version of the offshore database.

Table 1. Offshore wind resource area and potential by wind speed interval and state within 50 nm of shore.

	Wind Speed (m/s) at 90 m							
	7.0 - 7.5	7.5 - 8.0	8.0 - 8.5	8.5 - 9.0	9.0 - 9.5	9.5 - 10.0	>10.0	Total >7.0
State	km² (GW)	km² (GW)	km² (GW)	km² (GW)	km² (GW)	km² (GW)	km² (GW)	km² (GW)
California	11,439 (57.2)	24,864 (124.3)	23,059 (115.3)	22,852 (114.3)	13,185 (65.9)	15,231 (76.2)	6,926 (34.6)	117,555 (587.8)
Connecticut	530 (2.7)	702 (3.5)	40 (0.2)	0 (0.0)	0 (0.0)	0 (0.0)	0 (0.0)	1,272 (6.4)
Delaware	223 (1.1)	724 (3.6)	1,062 (5.3)	931 (4.7)	0 (0.0)	0 (0.0)	0 (0.0)	2,940 (14.7)
Georgia	3,820 (19.1)	7,741 (38.7)	523 (2.6)	0 (0.0)	0 (0.0)	0 (0.0)	0 (0.0)	12,085 (60.4)
Hawaii	18,873 (94.4)	42,298 (211.5)	33,042 (165.2)	13,913 (69.6)	7,779 (38.9)	6,720 (33.6)	4,852 (24.3)	127,477 (637.4)
Illinois	92 (0.5)	166 (0.8)	3,844 (19.2)	90 (0.4)	0 (0.0)	0 (0.0)	0 (0.0)	4,192 (21.0)
Indiana	82 (0.4)	216 (1.1)	286 (1.4)	0 (0.0)	0 (0.0)	0 (0.0)	0 (0.0)	584 (2.9)
Louisiana	48,043 (240.2)	15,032 (75.2)	0 (0.0)	0 (0.0)	0 (0.0)	0 (0.0)	0 (0.0)	63,075 (315.4)
Maine	906 (4.5)	1,142 (5.7)	1,976 (9.9)	3,331 (16.7)	8,429 (42.1)	15,485 (77.4)	42 (0.2)	31,311 (156.6)
Maryland	2,192 (11.0)	1,946 (9.7)	1,540 (7.7)	5,078 (25.4)	0 (0.0)	0 (0.0)	0 (0.0)	10,756 (53.8)
Massachusetts	202 (1.0)	526 (2.6)	1,639 (8.2)	3,606 (18.0)	20,351 (101.8)	13,674 (68.4)	0 (0.0)	39,997 (200.0)
Michigan	4,459 (22.3)	18,074 (90.4)	31,086 (155.4)	34,305 (171.5)	8,719 (43.6)	0 (0.0)	0 (0.0)	96,642 (483.2)
Minnesota	3,102 (15.5)	994 (5.0)	0 (0.0)	0 (0.0)	0 (0.0)	0 (0.0)	0 (0.0)	4,096 (20.5)
New Hampshire	19 (0.1)	46 (0.2)	171 (0.9)	336 (1.7)	102 (0.5)	0 (0.0)	0 (0.0)	672 (3.4)
New Jersey	528 (2.6)	1,508 (7.5)	4,965 (24.8)	12,934 (64.7)	0 (0.0)	0 (0.0)	0 (0.0)	19,935 (99.7)
New York	1,105 (5.5)	4,358 (21.8)	8,324 (41.6)	2,876 (14.4)	7,453 (37.3)	5,322 (26.6)	0 (0.0)	29,439 (147.2)
North Carolina	1,847 (9.2)	4,098 (20.5)	13,655 (68.3)	39,875 (199.4)	16 (0.1)	0 (0.0)	0 (0.0)	59,491 (297.5)
Ohio	341 (1.7)	3,067 (15.3)	5,829 (29.1)	0 (0.0)	0 (0.0)	0 (0.0)	0 (0.0)	9,237 (46.2)
Oregon	388 (1.9)	1,493 (7.5)	8,644 (43.2)	13,925 (69.6)	7,394 (37.0)	6,065 (30.3)	5,986 (29.9)	43,894 (219.5)

Table 1 (continued) Offshore wind resource area and potential by wind speed interval and state within 50 nm of shore.

	Wind Speed (m/s) at 90 m							
	7.0 - 7.5	7.5 - 8.0	8.0 - 8.5	8.5 - 9.0	9.0 - 9.5	9.5 - 10.0	>10.0	Total >7.0
Pennsylvania	34	211	1,679	0	0	0	0	1,924
	(0.2)	*(1.1)*	*(8.4)*	*(0.0)*	*(0.0)*	*(0.0)*	*(0.0)*	*(9.6)*
Rhode Island	224	126	283	671	1,461	2,360	0	5,126
	(1.1)	*(0.6)*	*(1.4)*	*(3.4)*	*(7.3)*	*(11.8)*	*(0.0)*	*(25.6)*
South Carolina	1,457	8,202	10,384	6,007	0	0	0	26,049
	(7.3)	*(41.0)*	*(51.9)*	*(30.0)*	*(0.0)*	*(0.0)*	*(0.0)*	*(130.2)*
Texas	2,019	24,823	16,556	12,273	0	0	0	55,671
	(10.1)	*(124.1)*	*(82.8)*	*(61.4)*	*(0.0)*	*(0.0)*	*(0.0)*	*(278.4)*
Virginia	889	3,658	6,549	7,794	0	0	0	18,890
	(4.4)	*(18.3)*	*(32.7)*	*(39.0)*	*(0.0)*	*(0.0)*	*(0.0)*	*(94.4)*
Washington	1,573	4,621	18,261	0	0	0	0	24,455
	(7.9)	*(23.1)*	*(91.3)*	*(0.0)*	*(0.0)*	*(0.0)*	*(0.0)*	*(122.3)*
Wisconsin	3,715	3,405	7,761	8,417	0	0	0	23,298
	(18.6)	*(17.0)*	*(38.8)*	*(42.1)*	*(0.0)*	*(0.0)*	*(0.0)*	*(116.5)*
Total	108,102	174,040	201,159	189,213	74,888	64,856	17,805	830,064
	(540.5)	*(870.2)*	*(1,005.8)*	*(946.1)*	*(374.4)*	*(324.3)*	*(89.0)*	*(4,150.3)*

Offshore Wind Resource

In May 2008, the U.S. Department of Energy (DOE) released a report detailing a deployment scenario by which the United States could achieve 20% of its electric energy supply from wind energy (U.S. Department of Energy 2008). Under this scenario, offshore wind was an essential contributor, providing 54 gigawatts of installed electric capacity to the grid. When President Obama took office in January 2009, his message clearly reinforced this challenge in a broader context of energy independence, environmental stewardship, and a strengthened economy based on clean renewable energy sources.

To achieve the deployment levels described in the 20% wind report, many technical and economic challenges must be faced. Many coastal areas in the United States have large electricity demand but have limited access to a high-quality land-based wind resource, and these areas are typically limited in their access to interstate grid transmission. Offshore wind resources have the potential to be a significant domestic renewable energy source for coastal electricity loads. The development of a reference and validated offshore wind resource database is one of the first steps necessary to understand the magnitude of the resource and to plan the distribution and development of future offshore wind power facilities.

DOE and the National Renewable Energy Laboratory (NREL) are working to assess the full potential of the nation's indigenous wind energy resources by creating a validated, national database that defines the significant characteristics used to quantify resource availability and its distribution. These elements include level of resource (annual average wind speed), water depth, distance from shore, and state administrative areas. The database will be periodically revised to reflect better wind resource estimates and to include updated information from other datasets.

This database serves as the foundation for future modifications that may include specific exclusion areas for the calculation of offshore wind resource potential. Earlier estimates of offshore wind potential in the United States made conservative assumptions to exclude large areas from development, based on their distance from shore (Musial and Butterfield 2004). In the 2004 study, all wind resources within 5 nautical miles (nm) of shore were excluded from development, two-thirds of the resource, 5 nm to 20 nm from shore, was excluded, and one-third of the resource was excluded for those areas greater than 20 nm from shore.

The paper by Musial and Butterfield detailed the rationale for exclusion factors but did not contain a detailed analysis of this subject. The basis for exclusion is a complex issue that should involve multiple stakeholders and negotiations. The development of offshore resources will be influenced by environmental and other ocean-use factors not currently included in the wind resource database.

Therefore, the current standard database quantifies the gross offshore wind resource. Exclusions from offshore development, of which there will be many, must be done on a state or regional basis to assure that local issues are addressed properly such as was attempted for areas along the Atlantic coast (Dhanju et. al 2008, Applied Technology and Management 2007).

The database provides the core information useful for the planning of resource-based offshore renewable energy development. The database incorporates the best offshore wind resource estimates with parameters that have a significant impact on offshore resource development. The main parameters include annual average wind speed, water depth (U.S. Department of Commerce's National Oceanic and Atmospheric Administration's [NOAA] Coastal Relief Model), distance from shore (shoreline delineation by NOAA), and administrative jurisdiction (U.S. Department of the Interior's Minerals Management Service [MMS]). As previously mentioned, the database was developed using Geographic Information Systems (GIS) techniques, allowing for spatial correlation of these characteristics. The offshore information presented on a state-by-state basis in this report is the first published version of the NREL offshore database. The database will be updated periodically as new offshore wind resource studies are completed and/or new information layers become available that directly impact offshore development.

GIS Database Structure

A GIS database was chosen to house the offshore resource data because the datasets have a significant spatial component. All of the component datasets are spatially referenced to the same spatial base allowing rapid indexing of the different datasets to each other. A database user may compare information from different datasets in the same geographic location. The GIS database also allows portions of a dataset to be quickly updated as new information becomes available. The database is sufficiently flexible to allow new elements such as environmental exclusion areas and shipping lanes/navigation zones, for example, to be included in future versions. The horizontal resolution of the database grid cells is 100 m by 100 m. The database extends out from the shoreline to 50 nm and includes major bays and inlets. Exceptions to the 50 nm mapped distance are the Great Lakes, where the entire lakes were mapped for the offshore resource, and Massachusetts, where the computed resource did not extend 50 nm from the edge of Nantucket Island and Martha's Vineyard in southeastern Massachusetts.

For ease of use and maintenance, the database has been divided into four regions initially: 1) Pacific coast (California, Oregon and Washington); 2) Atlantic coast (Maine, New Hampshire, Massachusetts, Rhode Island, Connecticut, New York, New Jersey, Delaware, Maryland, Virginia, North Carolina, South Carolina, Georgia, and Florida); 3) Gulf of Mexico (Florida, Alabama, Mississippi, Louisiana, and Texas); and 4) Great Lakes (New York, Pennsylvania, Ohio, Michigan, Indiana, Illinois, Wisconsin, and Minnesota). Hawaii, Alaska, and U.S. territories are handled separately.

Database Components

The database contains wind resource information in two types of fields. The first field is wind power class introduced in the "Wind Energy Resource Atlas of the United States" (Elliott et. al. 1987) at a 50 m height above the surface. The second field is the annual average wind speed at 90 m above the surface, the approximate hub-height of many current-day offshore wind turbines. This report presents information from this second field in the maps, figures, and tables in the body and appendices. The annual speeds in the database are binned in 0.25 meters per second (m/s) intervals, with the speed in the database defined as the midpoint of the interval. For example, a grid cell with an annual speed between 7.0 and 7.25 m/s receives a database value of 7.125 m/s. The speeds (and power classes) are binned because the elements in the database are required to be represented in discrete intervals. The database indexes wind resource potential by water depth at intervals of 10 m, distance from shore out to 50 nm in increments of 1 nm, and the best available offshore state and federal administrative boundaries, provided primarily by the MMS. Indexing allows the database to be summarized in different ways using these attributes (e.g. resource within 0-5 nm or 0-3 nm of shoreline).

Wind Resource Estimates

Annual average wind speeds are closely related to the available energy at a particular location and are categorized in the database by their value at a height of 90 m above the surface. The offshore wind resources of the United States were first estimated by NREL in 2003 (Musial and Butterfield 2004). Since then, updated offshore wind mapping projects (e.g. Elliott and Schwartz 2006) are gradually being completed. The updated maps provide a better estimate of the offshore wind resource than was previously available. At present, updated maps are available for the offshore areas of Georgia, Texas/Louisiana, northern New England (Massachusetts, New Hampshire, and Maine) and the states bordering the Great Lakes.

The updated wind resource maps were produced using a physics-based numerical computer model that provided preliminary estimates of the annual average wind resource. The modeling was developed by AWS Truepower (AWST) of Albany, New York, under subcontract to NREL, using their proprietary MesoMap system. The horizontal resolution of the model output is 200 m. The preliminary model estimates were validated by NREL using data from a variety of sources including ocean buoys, marine automated stations, Coast Guard stations and lighthouses, and satellite-derived 10 m wind speeds over the ocean estimated from the "state of the sea" as measured by microwave imaging. The wind measurements from the stations or grid points (in the case of satellite measurements) were extrapolated to 50 m above the surface and compared to the model estimates at the same height using sheer exponents from the Power Law Equation (Elliott et. al. 1987). The results of the validation model-measurement comparison were assembled into a spreadsheet and reviewed by AWST. In addition, NREL also produced qualitative comments on the validation results including recommended modifications to the preliminary resource estimates. Final modifications were agreed to after consultations between NREL and AWST. AWST adjusted the model output to reflect the modifications. There were not sufficient tall tower data to perform a high-quality validation at 90 m. Therefore, the modifications to the preliminary 90 m wind speed model output were based on the 50 m validation results. This adds some uncertainty to the final potential estimates, but should not significantly affect the scope of the offshore potential. NREL converted the final data into wind resource maps. The wind resource data were re-sampled from 200 m to 100 m horizontal resolution for inclusion in the database.

Calculations of the offshore wind resource estimates for states with older offshore wind maps (originally completed as part of an onshore wind mapping project) were done in two ways. For eight states, data from an older non-validated preliminary (model estimate) offshore wind map and a validated older final map were combined to calculate the offshore resource. Preliminary non-validated maps were used because: 1) older final wind maps extended only 5-10 nm offshore, while the preliminary maps extended further off the coast, thus reducing the area extrapolated to 50 nm; and 2) using the preliminary data resulted in more realistic offshore wind speed gradients than the gradients produced by extrapolating the older final map values from their seaward edges out to an additional 40-45 nm to the 50 nm line. Older final maps for five states (including the Atlantic coast of New York) extended either to the 50 nm line or the state boundary. In these cases, the data from the older final map was used to calculate the offshore potential without extrapolation. The complete listing of the older and updated offshore maps used for this project is in Appendix A. The states of Florida, Alabama, and Mississippi did not have any preliminary or final wind maps available. The lack of tall tower wind measurement data and offshore wind maps for this region made an estimate of the 90 m wind speeds problematic. Therefore, the offshore wind resource for these states was not included in this report. They will be included in the database once updated offshore maps for these states are complete.

The 90 m average wind speeds were calculated in several ways depending on the available height of the older preliminary and final map data. The states along the Atlantic coast from Rhode Island to North Carolina, and the state of Hawaii had offshore map wind speed values at 70 m and 100 m above the surface. For these states, the 90 m wind speed was calculated by a linear interpolation between the 70 m and 100 m wind speeds. The states of South Carolina, Washington, and Oregon had only 50 m map wind speeds available. The 90 m wind speeds for these states were calculated using a power law wind speed shear exponent (Elliott et. al. 1987) of 0.11. This exponent value was chosen based on the validation experience with the updated offshore wind maps and because other analyses of offshore wind resources indicate that the shear exponent is most often in the range from 0.08 to 0.14 for the offshore regions of the United States. The wind speeds at 90 m were about 6.5% higher than the 50 m wind speeds using the 0.11 shear exponent. Wind speed maps at 50 m and 70 m were available for California. The 90 m speed off the California coast was calculated assuming the speed shear exponent calculated between the 50 m and 70 m levels was also valid for the wind speeds between 70 m and 90 m.

For eight states (the Pacific coast and the Atlantic Coast from New Jersey to North Carolina), where both the older preliminary and final maps were used to estimate the wind resource, a blend of the calculated 90 m speeds from the preliminary and final maps was used to extrapolate the offshore potential to 50 nm. Speed values from the final maps were used to calculate the wind resource from close to shore out to the boundary between the final and preliminary maps. The speed at this boundary was calculated as a blend of the two final and preliminary map values. The preliminary map wind speeds were used from this boundary to the seaward edge of the preliminary map data. The wind speed values at the seaward edge of the preliminary map were held constant and expanded to the 50 nm line. The extrapolation from the edge of the preliminary map is a source of uncertainty in the final results for these eight states.

Horizontal discontinuities (seams) in the wind resource are present at several state boundaries. The discontinuities result from offshore data that is based on different versions of the numerical model used for the different onshore mapping projects and the way the extrapolation software interprets the data on either side of the seam. The most prominent seam appears near the border of Oregon and California. The resulting wind speed gradients in that region are not realistic. Other noticeable seams are located on the borders of New York and New Jersey and North Carolina and South Carolina. These seams in the maps and data interpretations further demonstrate the need to complete the updated offshore wind resource maps along the Atlantic and Pacific coasts.

Individual state and regional datasets were combined to form a composite image, based on 2009 map estimates, of the national offshore wind resource (Figure 1). The datasets are listed in Appendix A. The wind resource information is categorized by wind speed. However, for this initial report only areas with annual average wind speeds of 7.0 m/s and greater are included in the wind potential estimates. Economic factors make development of areas with less than 7.0 m/s average wind speeds unlikely. This delineation may be adjusted in future versions of the database as warranted. Examinations of the offshore wind resource distribution show an abundant wind resource pool, with wind resources greater than 7.0 m/s, located in many offshore areas of the country.

Bathymetry

The depth of the water affects the type of technology used to develop a given offshore wind resource project (Musial 2007). Current offshore wind turbine technology uses monopoles and gravity foundations in shallow water (0 m to 30 m). In transitional depths (30 m to 60 m), tripods, jackets and truss-type towers will be used. Deep water (> 60 m depth) may require floating structures instead of fixed bottom foundations, but this technology is currently in an early stage of development.

The bathymetry data used in the database has been categorized into 10 m water depth intervals, with depth calculated relative to the mean water surface height (Figure 2). The majority of the data was acquired from the U.S. Department of Commerce National Oceanic and Atmospheric Administration (NOAA) Coastal Relief Model. This dataset has a spatial resolution of 90 m. However, Lake Superior and some coastal areas further from shore were not included in the NOAA dataset. For those areas, bathymetry was interpolated from a spatially coarser global NOAA bathymetry dataset.

The East coast and the Gulf of Mexico have extensive areas of shallow water relatively far from shore. On the West coast, the continental shelf descends rapidly into the deep water category. The water depth also increases rapidly away from shore around Hawaii. In the Great Lakes region, Lake Erie and portions of Lake Ontario can be characterized as shallow; the other lakes are primarily deep water, with narrow bands of shallow and transitional water near the shore.

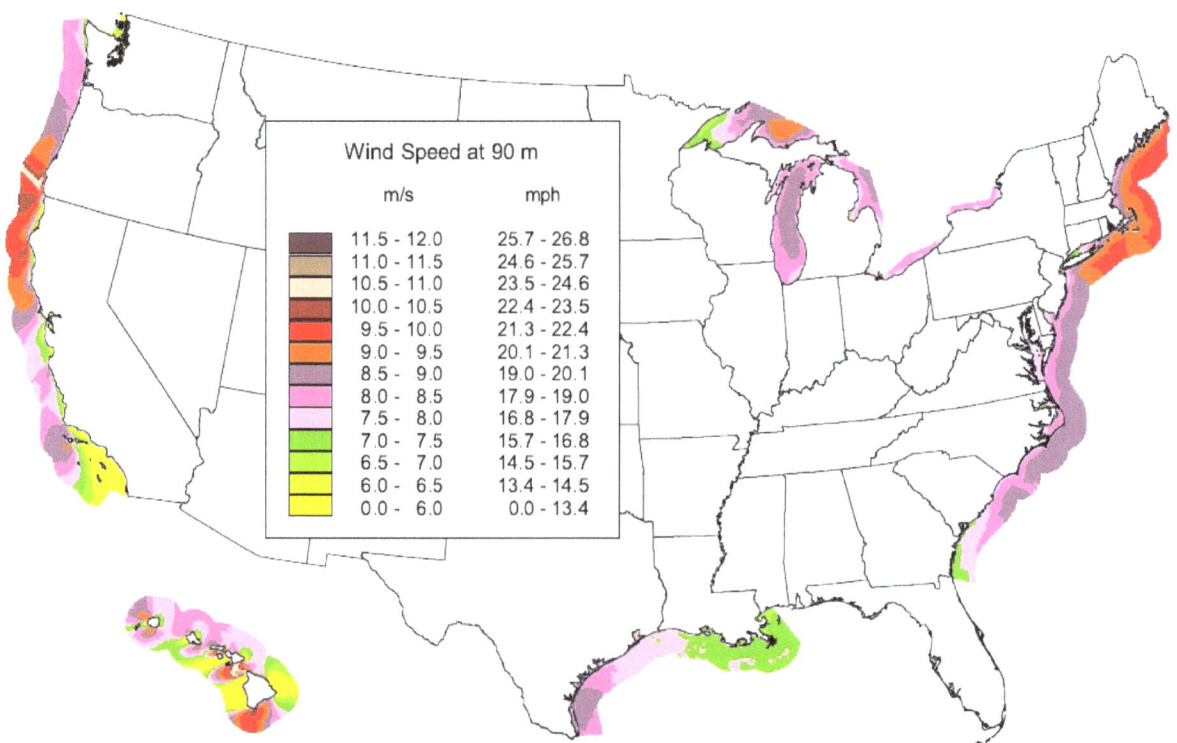

Figure 1. United States offshore wind resource at 90 m above the surface.

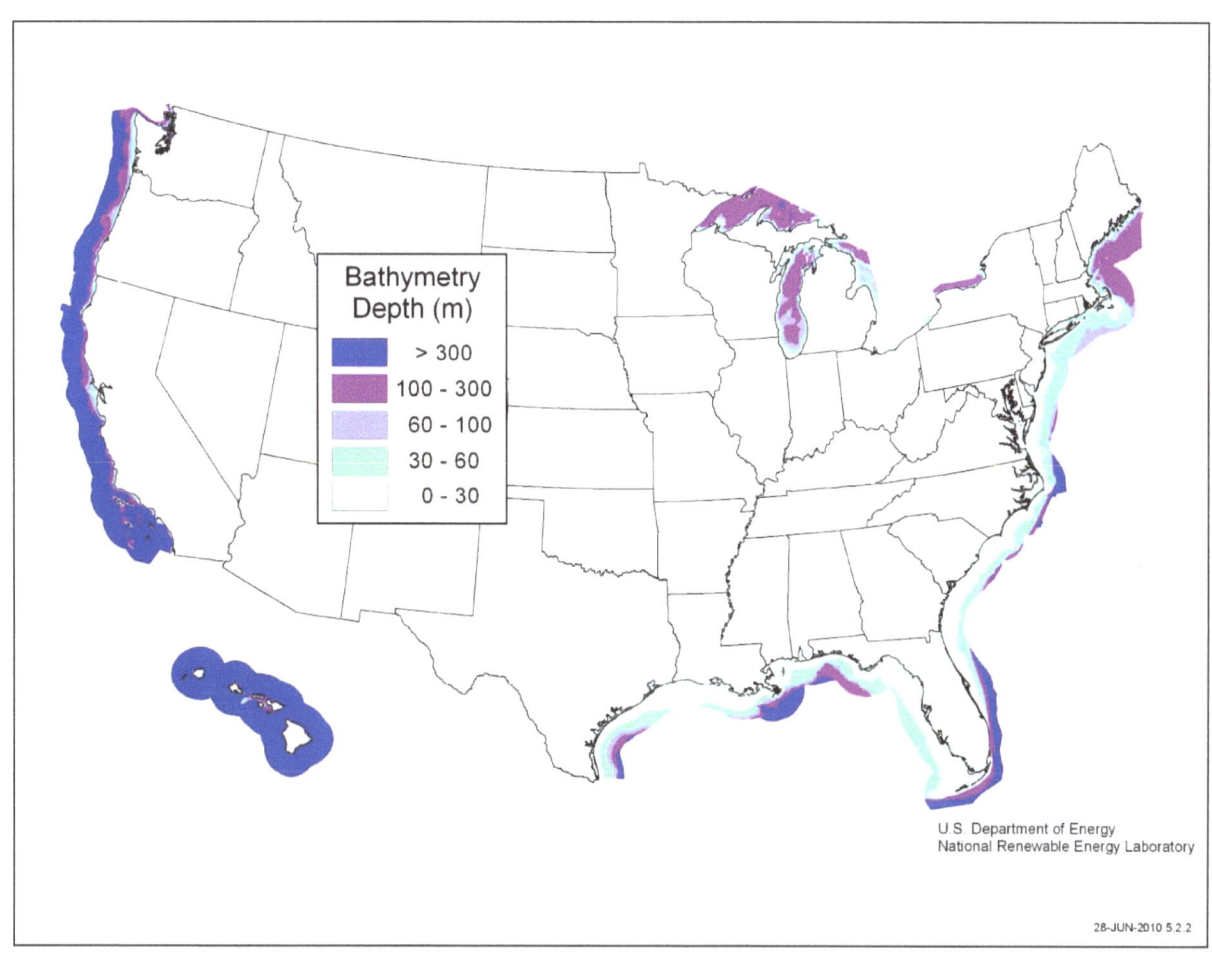

Figure 2. United States bathymetry distribution.

Distance From Shore

The distance a wind project is from shore determines a project's visibility from shore, and whether it is located in state or federal jurisdiction. Distance affects the potential cost of development through considerations such as the length of underwater cable needed to connect the offshore wind project to land-based electricity distribution facilities. In addition, coastline definition is complex because it is derived from a series of baseline points representing the mean lower low water line in direct contact with the open ocean (Thormahlen 1999). Some of these points can be seaward of the contiguous shoreline and change over time due to accretion and erosion of the shoreline. The MMS computed geographic lines from these points determining the boundary of state/federal offshore jurisdictions required by the Submerged Lands Act (SLA) (Thormahlen 1999). In Hawaii, the state/federal boundary was determined by NOAA. The database uses nm in its distance calculations. Federal jurisdiction begins 3 nm from the MMS baseline, except for Texas and the Gulf coast of Florida, where it begins at 9 nm. For the Atlantic coast, Pacific coast, and Gulf of Mexico distances are measured from the SLA line towards or away from shore, extending seaward a maximum of 50 nm. For the Great Lakes and Hawaii, distances were measured from the shoreline (Figure 3).

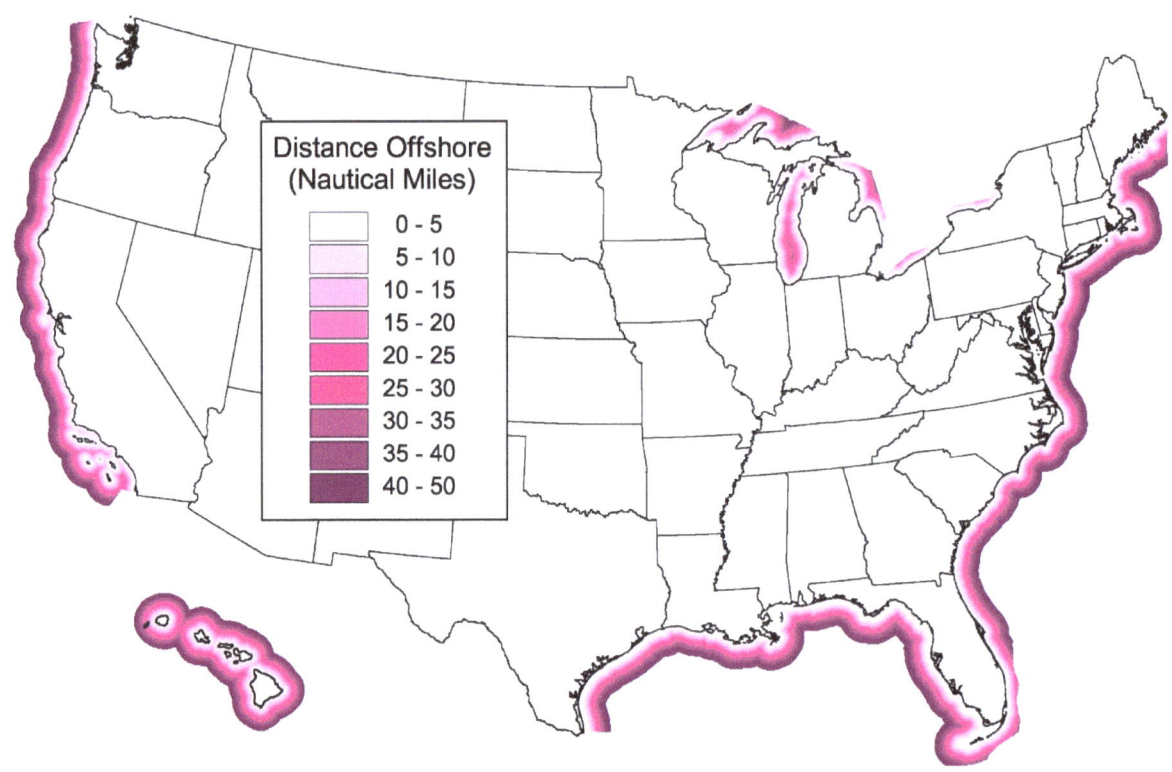

Figure 3. Distance from United States shoreline (nm).

Offshore Administrative Units

The determination of offshore jurisdiction encompasses complex legal agreements between individual states, legal agreements between the individual states and the federal government, and treaties between the United States and adjacent countries. Some of these boundaries are currently unresolved (New Jersey v. Delaware, Supreme Court Decision No. 134 Original, October Term 2007, and Thormahlen 1999). The state/federal offshore boundary is determined by the SLA and individual Supreme Court decisions for Texas and Florida (Thormahlen 1999). Seaward of the SLA, the MMS has proposed state boundaries that extend from the SLA line to the limit of the United States Outer Continental Shelf (OCS) based on the United Nations Convention on the Law of the Sea (Federal Register). Landward of the SLA line, state boundaries are based on legal agreements dating back to the Colonial period. A national dataset is still under development by the MMS and NOAA. For this report, NREL constructed an offshore administrative boundaries dataset from the MMS SLA, OCS, and OCS Administrative Boundaries, and individual state and local government administrative boundary datasets (Figure 4, by color). Where there was no available state data landward of the SLA, NREL constructed lines from the SLA to the shoreline. The list of data sources is provided in Appendix A.

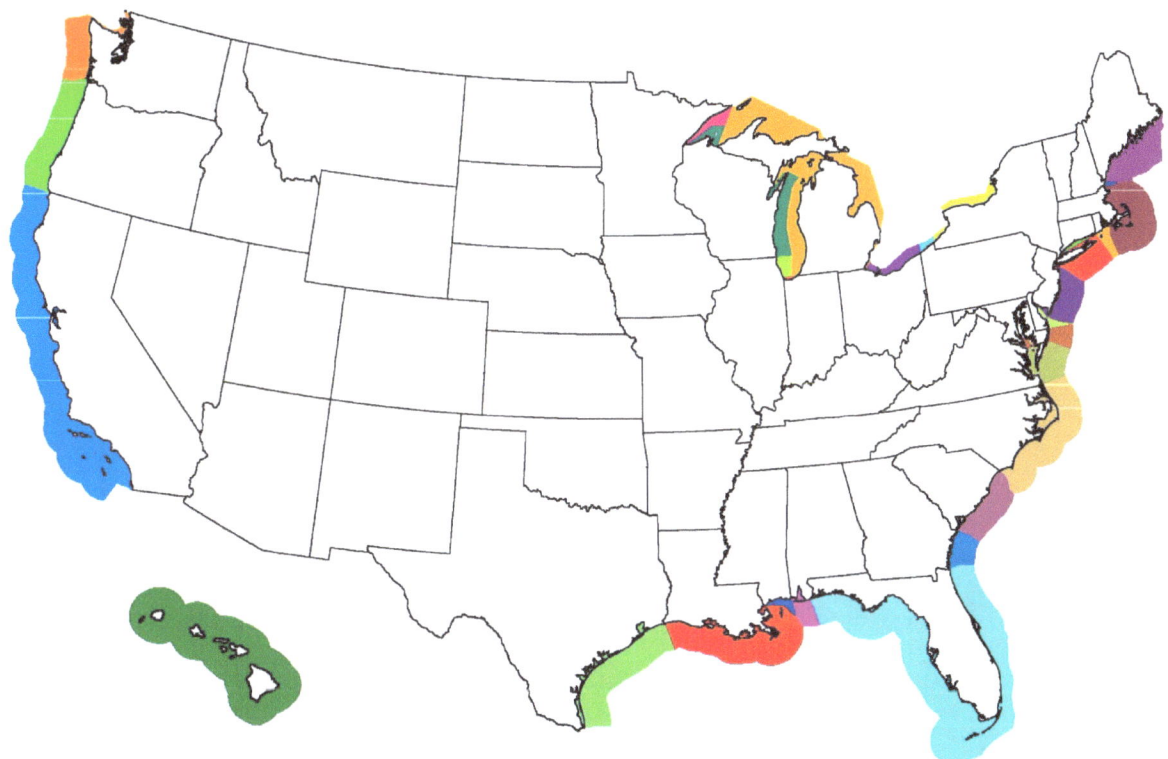

Figure 4. State administrative areas.

Database Summary

The offshore wind resource database contains the wind resource, water depth, distance from shore, and administrative unit for each 100 m by 100 m area out to a distance of 50 nm. The data consist of the gross wind energy resources without considering any exclusion areas. In other words, these tables do not exclude the wind resource as a function of distance from shore (Musial and Butterfield 2004), nor do they exclude the available resource due to environmental and other ocean-use factors. These data are the best available offshore wind resource data at present. The uncertainty of these resource estimates will significantly decrease as additional updated maps are completed.

Table 2 shows the gross wind resource at 90 m by state. The available square kilometers (km^2) of water and potential installed capacity in gigawatts (GW) is presented by wind speed interval (for annual wind speeds 7.0 m/s and higher), water depth, and distance from shore. A uniform factor of 5 megawatts/km^2 was applied to calculate the potential installed capacity. Detailed information by database element for each state is presented in Appendix B.

Table 2. Offshore wind resource area by state with potential by wind speed interval, water depth, distance from shore

State	Wind Speed at 90m m/s	0 - 3 nm Depth Category (m) 0 - 30 km² (GW)	0 - 3 nm 30 - 60 km² (GW)	0 - 3 nm > 60 km² (GW)	3 - 12 nm Depth Category (m) 0 - 30 km² (GW)	3 - 12 nm 30 - 60 km² (GW)	3 - 12 nm > 60 km² (GW)	12 - 50 nm Depth Category (m) 0 - 30 km² (GW)	12 - 50 nm 30 - 60 km² (GW)	12 - 50 nm > 60 km² (GW)	Total km² (GW)
California	7.0-7.5	266.2 (1.3)	236.2 (1.2)	257.4 (1.3)	100.9 (0.5)	456.8 (2.3)	4,554.0 (22.8)	7.7 (0.0)	22.9 (0.1)	5,536.8 (27.7)	11,438.9 (57.2)
	7.5-8.0	239.1 (1.2)	256.9 (1.3)	189.6 (0.9)	78.8 (0.4)	595.7 (3.0)	3,854.6 (19.3)	0.0 (0.0)	32.9 (0.2)	19,616.1 (98.1)	24,863.7 (124.3)
	8.0-8.5	125.2 (0.6)	178.2 (0.9)	281.8 (1.4)	7.1 (0.0)	105.8 (0.5)	4,539.1 (22.7)	0.0 (0.0)	0.0 (0.0)	17,822.2 (89.1)	23,059.3 (115.3)
	8.5-9.0	43.2 (0.2)	141.7 (0.7)	176.4 (0.9)	0.6 (0.0)	38.0 (0.2)	4,559.8 (22.8)	0.0 (0.0)	0.0 (0.0)	17,892.0 (89.5)	22,851.6 (114.3)
	9.0-9.5	2.1 (0.0)	18.8 (0.1)	14.9 (0.1)	0.0 (0.0)	0.9 (0.0)	988.0 (4.9)	0.0 (0.0)	0.0 (0.0)	12,160.2 (60.8)	13,184.8 (65.9)
	9.5-10.0	0.0 (0.0)	6.0 (0.0)	13.9 (0.1)	0.0 (0.0)	0.0 (0.0)	656.1 (3.3)	0.0 (0.0)	0.0 (0.0)	14,554.8 (72.8)	15,230.7 (76.2)
	>10.0	0.0 (0.0)	0.0 (0.0)	0.1 (0.0)	0.0 (0.0)	0.0 (0.0)	288.1 (1.4)	0.0 (0.0)	0.0 (0.0)	6,637.6 (33.2)	6,925.8 (34.6)
Connecticut	7.0-7.5	500.1 (2.5)	30.2 (0.2)	0.0 (0.0)	0.0 (0.0)	0.0 (0.0)	0.0 (0.0)	0.0 (0.0)	0.0 (0.0)	0.0 (0.0)	530.4 (2.7)
	7.5-8.0	617.3 (3.1)	83.0 (0.4)	1.2 (0.0)	0.0 (0.0)	0.0 (0.0)	0.0 (0.0)	0.0 (0.0)	0.0 (0.0)	0.0 (0.0)	701.6 (3.5)
	8.0-8.5	34.6 (0.2)	5.0 (0.0)	0.4 (0.0)	0.0 (0.0)	0.0 (0.0)	0.0 (0.0)	0.0 (0.0)	0.0 (0.0)	0.0 (0.0)	40.1 (0.2)
Delaware	7.0-7.5	223.2 (1.1)	0.0 (0.0)	0.0 (0.0)	0.0 (0.0)	0.0 (0.0)	0.0 (0.0)	0.0 (0.0)	0.0 (0.0)	0.0 (0.0)	223.2 (1.1)
	7.5-8.0	716.5 (3.6)	2.0 (0.0)	0.0 (0.0)	5.2 (0.0)	0.0 (0.0)	0.0 (0.0)	0.0 (0.0)	0.0 (0.0)	0.0 (0.0)	723.7 (3.6)
	8.0-8.5	135.4 (0.7)	10.5 (0.1)	0.0 (0.0)	658.0 (3.3)	8.6 (0.0)	0.0 (0.0)	240.4 (1.2)	8.9 (0.0)	0.0 (0.0)	1,061.9 (5.3)
	8.5-9.0	0.0 (0.0)	0.0 (0.0)	0.0 (0.0)	0.0 (0.0)	0.0 (0.0)	0.0 (0.0)	254.0 (1.3)	677.4 (3.4)	0.0 (0.0)	931.4 (4.7)

14

Table 2 (continued). Offshore wind resource area by state with potential by wind speed interval, water depth, distance from shore

State	Wind Speed at 90m m/s	Distance from Shoreline									Total	
		0 - 3 nm[1]			3 - 12 nm			12 - 50 nm				
		Depth Category (m)			Depth Category (m)			Depth Category (m)				
		0 - 30	30 - 60	> 60	0 - 30	30 - 60	> 60	0 - 30	30 - 60	> 60		
		km² (GW)	km² (GW)	km² (GW)	km² (GW)	km² (GW)	km² (GW)	km² (GW)	km² (GW)	km² (GW)	km² (GW)	
Georgia	7.0-7.5	547.5 (2.7)	0.0 (0.0)	0.0 (0.0)	2,162.3 (10.8)	0.0 (0.0)	0.0 (0.0)	1,110.6 (5.6)	0.0 (0.0)	0.0 (0.0)	3,820.4 (19.1)	
	7.5-8.0	85.2 (0.4)	0.0 (0.0)	0.0 (0.0)	529.6 (2.6)	0.0 (0.0)	0.0 (0.0)	5,204.3 (26.0)	1,922.1 (9.6)	0.0 (0.0)	7,741.2 (38.7)	
	8.0-8.5	0.0 (0.0)	0.0 (0.0)	0.0 (0.0)	0.0 (0.0)	0.0 (0.0)	0.0 (0.0)	3.9 (0.0)	519.5 (2.6)	0.0 (0.0)	523.4 (2.6)	
Hawaii	7.0-7.5	111.5 (0.6)	97.2 (0.5)	2,631.5 (13.2)	0.0 (0.0)	42.6 (0.2)	2,213.3 (11.1)	44.4 (0.2)	116.8 (0.6)	13,615.3 (68.1)	18,872.6 (94.4)	
	7.5-8.0	65.5 (0.3)	107.0 (0.5)	2,404.0 (12.0)	0.0 (0.0)	145.0 (0.7)	5,051.6 (25.3)	6.9 (0.0)	250.7 (1.3)	34,267.6 (171.3)	42,298.3 (211.5)	
	8.0-8.5	92.2 (0.5)	114.8 (0.6)	2,364.3 (11.8)	0.0 (0.0)	15.0 (0.1)	4,755.7 (23.8)	0.2 (0.0)	0.3 (0.0)	25,699.5 (128.5)	33,042.0 (165.2)	
	8.5-9.0	65.6 (0.3)	67.0 (0.3)	2,105.2 (10.5)	0.0 (0.0)	0.0 (0.0)	2,625.5 (13.1)	0.0 (0.0)	0.0 (0.0)	9,050.0 (45.3)	13,913.3 (69.6)	
	9.0-9.5	25.7 (0.1)	39.0 (0.2)	995.6 (5.0)	0.0 (0.0)	0.0 (0.0)	1,853.0 (9.3)	0.0 (0.0)	0.0 (0.0)	4,865.5 (24.3)	7,778.8 (38.9)	
	9.5-10.0	22.0 (0.1)	39.8 (0.2)	666.0 (3.3)	0.0 (0.0)	0.0 (0.0)	1,064.9 (5.3)	0.0 (0.0)	0.0 (0.0)	4,926.8 (24.6)	6,719.6 (33.6)	
	>10.0	26.3 (0.1)	70.5 (0.4)	1,344.0 (6.7)	0.0 (0.0)	0.0 (0.0)	2,027.6 (10.1)	0.0 (0.0)	0.0 (0.0)	1,383.5 (6.9)	4,852.0 (24.3)	
Illinois	7.0-7.5	91.6 (0.5)	0.0 (0.0)	0.0 (0.0)	0.0 (0.0)	0.0 (0.0)	0.0 (0.0)	0.0 (0.0)	0.0 (0.0)	0.0 (0.0)	91.6 (0.5)	
	7.5-8.0	164.6 (823.1)	0.0 (0.0)	0.0 (0.0)	0.0 (0.0)	0.1 (0.0)	0.0 (0.0)	0.0 (0.0)	1.3 (0.0)	0.0 (0.0)	166.0 (0.8)	
	8.0-8.5	244.2 (1.2)	1.0 (0.0)	0.0 (0.0)	830.4 (4.2)	512.2 (2.6)	140.7 (0.7)	7.1 (0.0)	426.1 (2.1)	1,682.8 (8.4)	3,844.5 (19.2)	
	8.5-9.0	0.0 (0.0)	0.0 (0.0)	0.0 (0.0)	13.4 (0.1)	76.6 (0.4)	0.0 (0.0)	0.0 (0.0)	0.0 (0.0)	0.0 (0.0)	89.9 (0.4)	

Table 2 (continued). Offshore wind resource area by state with potential by wind speed interval, water depth, distance from shore

State	Wind Speed at 90m m/s	Distance from Shoreline										
		0 - 3 nm[1]			3 - 12 nm			12 - 50 nm			Total	
		Depth Category (m)			Depth Category (m)			Depth Category (m)				
		0 - 30	30 - 60	> 60	0 - 30	30 - 60	> 60	0 - 30	30 - 60	> 60		
		km² (GW)	km² (GW)	km² (GW)	km² (GW)	km² (GW)	km² (GW)	km² (GW)	km² (GW)	km² (GW)	km² (GW)	
Indiana	7.0-7.5	82.1 (0.4)	0.0 (0.0)	0.0 (0.0)	0.0 (0.0)	0.0 (0.0)	0.0 (0.0)	0.0 (0.0)	0.0 (0.0)	0.0 (0.0)	82.1 (0.4)	
	7.5-8.0	153.8 (0.8)	0.0 (0.0)	0.0 (0.0)	62.6 (0.3)	0.0 (0.0)	0.0 (0.0)	0.0 (0.0)	0.0 (0.0)	0.0 (0.0)	216.3 (1.1)	
	8.0-8.5	101.5 (0.5)	0.0 (0.0)	0.0 (0.0)	184.0 (0.9)	0.0 (0.0)	0.0 (0.0)	0.0 (0.0)	0.0 (0.0)	0.0 (0.0)	285.5 (1.4)	
Louisiana	7.0-7.5	7,759.6 (38.8)	28.6 (0.1)	0.0 (0.0)	7,825.2 (39.1)	643.4 (3.2)	1,460.4 (7.3)	11,163.9 (55.8)	5,479.1 (27.4)	13,682.4 (68.4)	48,042.7 (240.2)	
	7.5-8.0	155.3 (0.8)	0.0 (0.0)	0.0 (0.0)	1,624.5 (8.1)	0.0 (0.0)	0.0 (0.0)	8,169.9 (40.8)	2,228.7 (11.1)	2,854.1 (14.3)	15,032.5 (75.2)	
Maine	7.0-7.5	787.0 (3.9)	91.2 (0.5)	11.9 (0.1)	7.8 (0.0)	4.8 (0.0)	3.5 (0.0)	0.0 (0.0)	0.0 (0.0)	0.0 (0.0)	906.2 (4.5)	
	7.5-8.0	797.2 (4.0)	285.4 (1.4)	19.4 (0.1)	6.7 (0.0)	19.6 (0.1)	14.1 (0.1)	0.0 (0.0)	0.0 (0.0)	0.0 (0.0)	1,142.3 (5.7)	
	8.0-8.5	777.0 (3.9)	440.8 (2.2)	74.2 (0.4)	63.4 (0.3)	385.6 (1.9)	234.5 (1.2)	0.0 (0.0)	0.0 (0.0)	0.0 (0.0)	1,975.6 (9.9)	
	8.5-9.0	513.4 (2.6)	614.0 (3.1)	157.6 (0.8)	18.2 (0.1)	219.1 (1.1)	1,401.9 (7.0)	0.0 (0.0)	0.0 (0.0)	406.8 (2.0)	3,331.1 (16.7)	
	9.0-9.5	142.2 (0.7)	390.0 (2.0)	309.2 (1.5)	25.9 (0.1)	469.0 (2.3)	3,504.1 (17.5)	0.0 (0.0)	57.8 (0.3)	3,530.9 (17.7)	8,429.2 (42.1)	
	9.5-10.0	5.5 (0.0)	24.9 (0.1)	42.3 (0.2)	1.0 (0.0)	38.3 (0.2)	1,459.8 (7.3)	0.0 (0.0)	7.4 (0.0)	13,905.6 (69.5)	15,484.7 (77.4)	
	>10.0	0.0 (0.0)	0.0 (0.0)	0.0 (0.0)	0.0 (0.0)	0.0 (0.0)	41.6 (0.2)	0.0 (0.0)	0.0 (0.0)	0.0 (0.0)	41.6 (0.2)	
Maryland	7.0-7.5	2,175.4 (10.9)	16.6 (0.1)	0.0 (0.0)	0.0 (0.0)	0.0 (0.0)	0.0 (0.0)	0.0 (0.0)	0.0 (0.0)	0.0 (0.0)	2,192.0 (11.0)	
	7.5-8.0	1,922.5 (9.6)	14.0 (0.1)	0.0 (0.0)	9.9 (0.0)	0.0 (0.0)	0.0 (0.0)	0.0 (0.0)	0.0 (0.0)	0.0 (0.0)	1,946.4 (9.7)	

16

Table 2 (continued). Offshore wind resource area by state with potential by wind speed interval, water depth, distance from shore

State	Wind Speed at 90m m/s	Distance from Shoreline									Total
		0 - 3 nm			3 - 12 nm			12 - 50 nm			
		Depth Category (m)			Depth Category (m)			Depth Category (m)			
		0 - 30 km² (GW)	30 - 60 km² (GW)	> 60 km² (GW)	0 - 30 km² (GW)	30 - 60 km² (GW)	> 60 km² (GW)	0 - 30 km² (GW)	30 - 60 km² (GW)	> 60 km² (GW)	km² (GW)
Maryland (cont.)	8.0-8.5	163.3 (0.8)	0.0 (0.0)	0.0 (0.0)	924.0 (4.6)	0.0 (0.0)	0.0 (0.0)	435.4 (2.2)	17.2 (0.1)	0.0 (0.0)	1,539.9 (7.7)
	8.5-9.0	0.0 (0.0)	0.0 (0.0)	0.0 (0.0)	0.0 (0.0)	0.0 (0.0)	0.0 (0.0)	531.3 (2.7)	3,210.7 (16.1)	1,336.1 (6.7)	5,078.1 (25.4)
Massachusetts	7.0-7.5	201.6 (1.0)	0.0 (0.0)	0.0 (0.0)	0.0 (0.0)	0.0 (0.0)	0.0 (0.0)	0.0 (0.0)	0.0 (0.0)	0.0 (0.0)	201.6 (1.0)
	7.5-8.0	521.4 (2.6)	4.7 (0.0)	0.0 (0.0)	0.0 (0.0)	0.0 (0.0)	0.0 (0.0)	0.0 (0.0)	0.0 (0.0)	0.0 (0.0)	526.1 (2.6)
	8.0-8.5	927.4 (4.6)	327.3 (1.6)	28.6 (0.1)	78.2 (0.4)	152.0 (0.8)	125.5 (0.6)	0.0 (0.0)	0.0 (0.0)	0.0 (0.0)	1,639.1 (8.2)
	8.5-9.0	1,508.2 (7.5)	378.1 (1.9)	12.6 (0.1)	315.0 (1.6)	354.5 (1.8)	812.2 (4.1)	11.4 (0.1)	23.5 (0.1)	190.4 (1.0)	3,606.0 (18.0)
	9.0-9.5	1,137.0 (5.7)	322.6 (1.6)	20.0 (0.1)	2,696.9 (13.5)	1,418.6 (7.1)	1,006.5 (5.0)	1,689.9 (8.4)	5,051.8 (25.3)	7,007.4 (35.0)	20,350.7 (101.8)
	9.5-10.0	2.0 (0.0)	0.0 (0.0)	0.0 (0.0)	8.6 (0.0)	119.2 (0.6)	0.0 (0.0)	472.1 (2.4)	3,459.5 (17.3)	9,612.5 (48.1)	13,674.0 (68.4)
Michigan	7.0-7.5	3,411.4 (17.1)	330.6 (1.7)	243.7 (1.2)	142.6 (0.7)	49.8 (0.2)	268.0 (1.3)	1.2 (0.0)	0.0 (0.0)	12.1 (0.1)	4,459.3 (22.3)
	7.5-8.0	4,632.7 (23.2)	1,258.2 (6.3)	635.3 (3.2)	3,069.0 (15.3)	2,916.4 (14.6)	2,991.0 (15.0)	14.2 (0.1)	108.5 (0.5)	2,448.4 (12.2)	18,073.7 (90.4)
	8.0-8.5	2,588.1 (12.9)	1,253.3 (6.3)	769.0 (3.8)	2,770.6 (13.9)	5,889.7 (29.4)	9,411.9 (47.1)	30.9 (0.2)	876.1 (4.4)	7,496.4 (37.5)	31,085.9 (155.4)
	8.5-9.0	138.7 (0.7)	156.2 (0.8)	396.5 (2.0)	145.1 (0.7)	550.6 (2.8)	8,580.8 (42.9)	118.5 (0.6)	1,383.9 (6.9)	22,834.5 (114.2)	34,304.7 (171.5)
	9.0-9.5	0.0 (0.0)	0.0 (0.0)	0.0 (0.0)	0.0 (0.0)	0.0 (0.0)	69.9 (0.3)	114.6 (0.6)	249.6 (1.2)	8,284.5 (41.4)	8,718.7 (43.6)

Table 2 (continued). Offshore wind resource area by state with potential by wind speed interval, water depth, distance from shore

State	Wind Speed at 90m m/s	Distance from Shoreline									
		0 - 3 nm			3 - 12 nm			12 - 50 nm			Total
		Depth Category (m)			Depth Category (m)			Depth Category (m)			
		0 - 30 km² (GW)	30 - 60 km² (GW)	> 60 km² (GW)	0 - 30 km² (GW)	30 - 60 km² (GW)	> 60 km² (GW)	0 - 30 km² (GW)	30 - 60 km² (GW)	> 60 km² (GW)	km² (GW)
Minnesota	7.0-7.5	5.1 (0.0)	43.6 (0.2)	272.6 (1.4)	0.0 (0.0)	31.5 (0.2)	1,718.1 (8.6)	0.0 (0.0)	0.0 (0.0)	1,031.1 (5.2)	3,101.9 (15.5)
	7.5-8.0	0.0 (0.0)	0.1 (0.0)	12.7 (0.1)	0.0 (0.0)	0.0 (0.0)	60.8 (0.3)	0.0 (0.0)	0.0 (0.0)	920.7 (4.6)	994.3 (5.0)
New Hampshire	7.0-7.5	18.6 (0.1)	0.0 (0.0)	0.0 (0.0)	0.0 (0.0)	0.0 (0.0)	0.0 (0.0)	0.0 (0.0)	0.0 (0.0)	0.0 (0.0)	18.6 (0.1)
	7.5-8.0	45.8 (0.2)	0.0 (0.0)	0.0 (0.0)	0.0 (0.0)	0.0 (0.0)	0.0 (0.0)	0.0 (0.0)	0.0 (0.0)	0.0 (0.0)	45.8 (0.2)
	8.0-8.5	44.6 (0.2)	29.7 (0.1)	0.0 (0.0)	6.9 (0.0)	75.5 (0.4)	14.0 (0.1)	0.0 (0.0)	0.0 (0.0)	0.0 (0.0)	170.6 (0.9)
	8.5-9.0	0.0 (0.0)	8.0 (0.0)	7.2 (0.0)	0.0 (0.0)	12.4 (0.1)	255.7 (1.3)	0.0 (0.0)	10.1 (0.1)	42.2 (0.2)	335.7 (1.7)
	9.0-9.5	0.0 (0.0)	0.0 (0.0)	0.0 (0.0)	0.0 (0.0)	0.0 (0.0)	0.0 (0.0)	0.0 (0.0)	35.2 (0.2)	66.4 (0.3)	101.6 (0.5)
New Jersey	7.0-7.5	520.2 (2.6)	0.0 (0.0)	0.0 (0.0)	8.2 (0.0)	0.0 (0.0)	0.0 (0.0)	0.0 (0.0)	0.0 (0.0)	0.0 (0.0)	528.5 (2.6)
	7.5-8.0	960.4 (4.8)	0.0 (0.0)	0.0 (0.0)	518.6 (2.6)	27.9 (0.1)	0.7 (0.0)	0.0 (0.0)	0.0 (0.0)	0.0 (0.0)	1,507.6 (7.5)
	8.0-8.5	776.9 (3.9)	0.0 (0.0)	0.0 (0.0)	2,764.8 (13.8)	127.1 (0.6)	3.2 (0.0)	899.4 (4.5)	372.4 (1.9)	21.6 (0.1)	4,965.4 (24.8)
	8.5-9.0	13.1 (0.1)	0.0 (0.0)	0.0 (0.0)	214.7 (1.1)	0.0 (0.0)	0.0 (0.0)	2,019.5 (10.1)	10,382.4 (51.9)	303.8 (1.5)	12,933.5 (64.7)
New York	7.0-7.5	1,047.9 (5.2)	54.2 (0.3)	0.0 (0.0)	2.8 (0.0)	0.0 (0.0)	0.0 (0.0)	0.0 (0.0)	0.0 (0.0)	0.0 (0.0)	1,104.8 (5.5)
	7.5-8.0	1,983.1 (9.9)	354.7 (1.8)	11.6 (0.1)	472.7 (2.4)	609.0 (3.0)	866.8 (4.3)	0.0 (0.0)	0.0 (0.0)	59.9 (0.3)	4,357.8 (21.8)
	8.0-8.5	1,329.3 (6.6)	602.9 (3.0)	196.5 (1.0)	375.6 (1.9)	535.7 (2.7)	3,122.0 (15.6)	42.2 (0.2)	12.6 (0.1)	2,107.4 (10.5)	8,324.1 (41.6)

Table 2 (continued). Offshore wind resource area by state with potential by wind speed interval, water depth, distance from shore

		Distance from Shoreline									
		0 - 3 nm			3 - 12 nm			12 - 50 nm			
		Depth Category (m)			Depth Category (m)			Depth Category (m)			Total
State	Wind Speed at 90m m/s	0 - 30	30 - 60	> 60	0 - 30	30 - 60	> 60	0 - 30	30 - 60	> 60	
		km² (GW)	km² (GW)	km² (GW)	km² (GW)	km² (GW)	km² (GW)	km² (GW)	km² (GW)	km² (GW)	km² (GW)
New York (cont.)	8.5-9.0	535.1 (2.7)	110.6 (0.6)	0.6 (0.0)	1,042.6 (5.2)	93.8 (0.5)	0.0 (0.0)	261.5 (1.3)	727.8 (3.6)	104.5 (0.5)	2,876.4 (14.4)
	9.0-9.5	5.4 (0.0)	0.0 (0.0)	0.0 (0.0)	391.9 (2.0)	1,407.6 (7.0)	0.0 (0.0)	6.4 (0.0)	4,926.1 (24.6)	716.0 (3.6)	7,453.3 (37.3)
	9.5-10.0	0.0 (0.0)	0.0 (0.0)	0.0 (0.0)	0.0 (0.0)	0.0 (0.0)	0.0 (0.0)	0.0 (0.0)	1,817.6 (9.1)	3,504.8 (17.5)	5,322.4 (26.6)
North Carolina	7.0-7.5	1,847.4 (9.2)	0.0 (0.0)	0.0 (0.0)	0.0 (0.0)	0.0 (0.0)	0.0 (0.0)	0.0 (0.0)	0.0 (0.0)	0.0 (0.0)	1,847.4 (9.2)
	7.5-8.0	3,002.1 (15.0)	0.0 (0.0)	0.0 (0.0)	1,096.1 (5.5)	0.0 (0.0)	0.0 (0.0)	0.0 (0.0)	0.0 (0.0)	0.0 (0.0)	4,098.3 (20.5)
	8.0-8.5	3,920.0 (19.6)	0.1 (0.0)	0.1 (0.0)	4,653.8 (23.3)	0.0 (0.0)	0.0 (0.0)	4,034.9 (20.2)	746.5 (3.7)	299.2 (1.5)	13,654.7 (68.3)
	8.5-9.0	105.0 (0.5)	0.0 (0.0)	0.0 (0.0)	2,964.1 (14.8)	261.4 (1.3)	8.9 (0.0)	6,382.4 (31.9)	14,519.6 (72.6)	15,633.3 (78.2)	39,874.8 (199.4)
	9.0-9.5	0.0 (0.0)	0.0 (0.0)	0.0 (0.0)	0.0 (0.0)	0.0 (0.0)	0.0 (0.0)	0.0 (0.0)	16.0 (0.1)	0.0 (0.0)	16.0 (0.1)
Ohio	7.0-7.5	341.0 (1.7)	0.0 (0.0)	0.0 (0.0)	0.1 (0.0)	0.0 (0.0)	0.0 (0.0)	0.0 (0.0)	0.0 (0.0)	0.0 (0.0)	341.0 (1.7)
	7.5-8.0	1,106.8 (5.5)	0.0 (0.0)	0.0 (0.0)	1,959.8 (9.8)	0.0 (0.0)	0.0 (0.0)	0.0 (0.0)	0.0 (0.0)	0.0 (0.0)	3,066.7 (15.3)
	8.0-8.5	565.2 (2.8)	0.0 (0.0)	0.0 (0.0)	2,193.1 (11.0)	0.0 (0.0)	0.0 (0.0)	3,071.2 (15.4)	0.0 (0.0)	0.0 (0.0)	5,829.5 (29.1)
Oregon	7.0-7.5	355.8 (1.8)	20.6 (0.1)	0.0 (0.0)	0.9 (0.0)	9.2 (0.0)	1.2 (0.0)	0.0 (0.0)	0.0 (0.0)	0.0 (0.0)	387.6 (1.9)
	7.5-8.0	523.0 (2.6)	319.2 (1.6)	37.6 (0.2)	46.5 (0.2)	231.8 (1.2)	335.0 (1.7)	0.0 (0.0)	0.0 (0.0)	0.0 (0.0)	1,493.0 (7.5)
	8.0-8.5	198.3 (1.0)	277.0 (1.4)	6.6 (0.0)	19.0 (0.1)	595.7 (3.0)	2,558.5 (12.8)	0.0 (0.0)	0.0 (0.0)	4,989.5 (24.9)	8,644.4 (43.2)

Table 2 (continued). Offshore wind resource area by state with potential by wind speed interval, water depth, distance from shore

		Distance from Shoreline									
		0 - 3 nm			3 - 12 nm			12 - 50 nm			
		Depth Category (m)			Depth Category (m)			Depth Category (m)			Total
State	Wind Speed at 90m m/s	0 - 30 km² (GW)	30 - 60 km² (GW)	> 60 km² (GW)	0 - 30 km² (GW)	30 - 60 km² (GW)	> 60 km² (GW)	0 - 30 km² (GW)	30 - 60 km² (GW)	> 60 km² (GW)	km² (GW)
Oregon (cont.)	8.5-9.0	64.1 (0.3)	98.8 (0.5)	0.6 (0.0)	0.0 (0.0)	108.0 (0.5)	1,967.2 (9.8)	0.0 (0.0)	45.6 (0.2)	11,640.3 (58.2)	13,924.6 (69.6)
	9.0-9.5	64.1 (0.3)	55.4 (0.3)	38.6 (0.2)	0.0 (0.0)	32.6 (0.2)	614.9 (3.1)	0.0 (0.0)	0.0 (0.0)	6,588.2 (32.9)	7,393.7 (37.0)
	9.5-10.0	47.3 (0.2)	80.4 (0.4)	14.6 (0.1)	0.0 (0.0)	33.8 (0.2)	634.6 (3.2)	0.0 (0.0)	0.0 (0.0)	5,254.6 (26.3)	6,065.3 (30.3)
	>10.0	0.2 (0.0)	19.4 (0.1)	33.3 (0.2)	0.0 (0.0)	18.2 (0.1)	1,368.5 (6.8)	0.0 (0.0)	0.0 (0.0)	4,546.0 (22.7)	5,985.5 (29.9)
Pennsylvania	7.0-7.5	34.2 (0.2)	0.0 (0.0)	0.0 (0.0)	0.0 (0.0)	0.0 (0.0)	0.0 (0.0)	0.0 (0.0)	0.0 (0.0)	0.0 (0.0)	34.2 (0.2)
	7.5-8.0	113.4 (0.6)	0.0 (0.0)	0.0 (0.0)	53.2 (0.3)	44.2 (0.2)	0.0 (0.0)	0.0 (0.0)	0.0 (0.0)	0.0 (0.0)	210.8 (1.1)
	8.0-8.5	276.2 (1.4)	0.9 (0.0)	0.0 (0.0)	750.8 (3.8)	358.1 (1.8)	0.0 (0.0)	222.9 (1.1)	70.2 (0.4)	0.0 (0.0)	1,679.1 (8.4)
Rhode Island	7.0-7.5	216.3 (1.1)	8.0 (0.0)	0.0 (0.0)	0.0 (0.0)	0.0 (0.0)	0.0 (0.0)	0.0 (0.0)	0.0 (0.0)	0.0 (0.0)	224.3 (1.1)
	7.5-8.0	123.2 (0.6)	2.8 (0.0)	0.0 (0.0)	0.0 (0.0)	0.0 (0.0)	0.0 (0.0)	0.0 (0.0)	0.0 (0.0)	0.0 (0.0)	126.0 (0.6)
	8.0-8.5	139.9 (0.7)	39.8 (0.2)	0.0 (0.0)	34.5 (0.2)	69.2 (0.3)	0.0 (0.0)	0.0 (0.0)	0.0 (0.0)	0.0 (0.0)	283.5 (1.4)
	8.5-9.0	120.3 (0.6)	93.0 (0.5)	0.0 (0.0)	183.6 (0.9)	274.2 (1.4)	0.0 (0.0)	0.0 (0.0)	0.0 (0.0)	0.0 (0.0)	671.2 (3.4)
	9.0-9.5	53.7 (0.3)	17.6 (0.1)	0.0 (0.0)	176.0 (0.9)	782.7 (3.9)	0.0 (0.0)	0.0 (0.0)	430.2 (2.2)	1.3 (0.0)	1,461.5 (7.3)
	9.5-10.0	0.0 (0.0)	0.0 (0.0)	0.0 (0.0)	0.0 (0.0)	5.8 (0.0)	0.0 (0.0)	0.0 (0.0)	967.2 (4.8)	1,386.8 (6.9)	2,359.8 (11.8)

Table 2 (continued). Offshore wind resource area by state with potential by wind speed interval, water depth, distance from shore

State	Wind Speed at 90m m/s	0 - 3 nm[1] Depth Category (m)			3 - 12 nm Depth Category (m)			12 - 50 nm Depth Category (m)			Total
		0 - 30 km² (GW)	30 - 60 km² (GW)	>60 km² (GW)	0 - 30 km² (GW)	30 - 60 km² (GW)	>60 km² (GW)	0 - 30 km² (GW)	30 - 60 km² (GW)	>60 km² (GW)	km² (GW)
South Carolina	7.0-7.5	848.2 (4.2)	0.0 (0.0)	0.0 (0.0)	608.4 (3.0)	0.0 (0.0)	0.0 (0.0)	0.0 (0.0)	0.0 (0.0)	0.0 (0.0)	1,456.5 (7.3)
	7.5-8.0	593.5 (3.0)	0.0 (0.0)	0.0 (0.0)	3,053.9 (15.3)	0.0 (0.0)	0.0 (0.0)	4,267.6 (21.3)	287.0 (1.4)	0.0 (0.0)	8,202.1 (41.0)
	8.0-8.5	22.9 (0.1)	0.0 (0.0)	0.0 (0.0)	1,609.4 (8.0)	0.0 (0.0)	0.0 (0.0)	4,151.4 (20.8)	3,925.6 (19.6)	674.5 (3.4)	10,383.7 (51.9)
	8.5-9.0	0.0 (0.0)	0.0 (0.0)	0.0 (0.0)	0.0 (0.0)	0.0 (0.0)	0.0 (0.0)	2,027.0 (10.1)	3,109.7 (15.5)	869.9 (4.3)	6,006.5 (30.0)
Texas[1]	7.0-7.5	1,785.5 (8.9)	0.0 (0.0)	0.1 (0.0)	96.2 (0.5)	0.0 (0.0)	0.0 (0.0)	137.2 (0.7)	0.0 (0.0)	0.0 (0.0)	2,019.1 (10.1)
	7.5-8.0	9,046.1 (45.2)	0.0 (0.0)	0.0 (0.0)	1,165.2 (5.8)	0.0 (0.0)	0.1 (0.0)	8,045.2 (40.2)	6,174.7 (30.9)	391.5 (2.0)	24,822.7 (124.1)
	8.0-8.5	5,928.5 (29.6)	125.4 (0.6)	0.0 (0.0)	742.1 (3.7)	136.9 (0.7)	0.0 (0.0)	583.5 (2.9)	4,442.8 (22.2)	4,597.1 (23.0)	16,556.3 (82.8)
	8.5-9.0	3,798.0 (19.0)	601.0 (3.0)	0.0 (0.0)	42.9 (0.2)	945.1 (4.7)	0.0 (0.0)	0.0 (0.0)	3,210.9 (16.1)	3,674.6 (18.4)	12,272.5 (61.4)
Virginia	7.0-7.5	889.2 (4.4)	0.0 (0.0)	0.0 (0.0)	0.0 (0.0)	0.0 (0.0)	0.0 (0.0)	0.0 (0.0)	0.0 (0.0)	0.0 (0.0)	889.2 (4.4)
	7.5-8.0	3,605.5 (18.0)	15.3 (0.1)	0.0 (0.0)	37.0 (0.2)	0.0 (0.0)	0.0 (0.0)	0.0 (0.0)	0.0 (0.0)	0.0 (0.0)	3,657.8 (18.3)
	8.0-8.5	1,135.8 (5.7)	2.0 (0.0)	0.0 (0.0)	2,985.9 (14.9)	0.0 (0.0)	0.0 (0.0)	2,355.9 (11.8)	69.1 (0.3)	0.0 (0.0)	6,548.7 (32.7)
	8.5-9.0	0.0 (0.0)	0.0 (0.0)	0.0 (0.0)	23.7 (0.1)	0.0 (0.0)	0.0 (0.0)	2,030.5 (10.2)	4,839.8 (24.2)	900.0 (4.5)	7,794.0 (39.0)

Distance from Shoreline

[1] Federal waters begin at 3 nm with the exception of Texas, which begins at 9 nm. The area reported for Texas represents 0-9 nm; 9-12 nm; and 12-50 nm respectively.

Table 2 (continued). Offshore wind resource area by state with potential by wind speed interval, water depth, distance from shore

State	Wind Speed at 90m m/s	Distance from Shoreline										Total
		0 - 3 nm[1]			3 - 12 nm			12 - 50 nm			
		Depth Category (m)			Depth Category (m)			Depth Category (m)			
		0 - 30	30 - 60	> 60	0 - 30	30 - 60	> 60	0 - 30	30 - 60	> 60	
		km² (GW)	km² (GW)	km² (GW)	km² (GW)	km² (GW)	km² (GW)	km² (GW)	km² (GW)	km² (GW)	km² (GW)
Washington	7.0-7.5	622.9 (3.1)	133.5 (0.7)	57.3 (0.3)	231.9 (1.2)	350.2 (1.8)	176.7 (0.9)	0.0 (0.0)	0.0 (0.0)	0.2 (0.0)	1,572.7 (7.9)
	7.5-8.0	370.6 (1.9)	0.0 (0.0)	0.0 (0.0)	203.0 (1.0)	841.6 (4.2)	1,173.0 (5.9)	0.0 (0.0)	0.0 (0.0)	2,033.1 (10.2)	4,621.4 (23.1)
	8.0-8.5	61.7 (0.3)	0.0 (0.0)	0.0 (0.0)	211.7 (1.1)	977.9 (4.9)	475.1 (2.4)	0.0 (0.0)	19.6 (0.1)	16,514.8 (82.6)	18,260.7 (91.3)
Wisconsin	7.0-7.5	1,053.8 (5.3)	268.2 (1.3)	145.8 (0.7)	349.9 (1.7)	616.6 (3.1)	1,085.0 (5.4)	0.0 (0.0)	34.2 (0.2)	162.0 (0.8)	3,715.4 (18.6)
	7.5-8.0	1,417.2 (7.1)	235.5 (1.2)	23.7 (0.1)	361.7 (1.8)	223.0 (1.1)	779.3 (3.9)	0.0 (0.0)	0.0 (0.0)	364.2 (1.8)	3,404.5 (17.0)
	8.0-8.5	830.6 (4.2)	372.7 (1.9)	0.0 (0.0)	451.4 (2.3)	1,386.1 (6.9)	2,539.0 (12.7)	0.0 (0.0)	21.3 (0.1)	2,160.1 (10.8)	7,761.1 (38.8)
	8.5-9.0	25.5 (0.1)	8.1 (0.0)	0.0 (0.0)	11.7 (0.1)	300.8 (1.5)	1,347.3 (6.7)	0.0 (0.0)	62.8 (0.3)	6,661.0 (33.3)	8,417.3 (42.1)

Future Plans

The data base is designed to be an evolving product with its elements subject to modification and change. The offshore wind resource is anticipated to undergo notable change as new updated resource maps are completed. The mid-Atlantic states, from Rhode Island to South Carolina, will be the next group of states to be mapped and validated. The database will be modified to include the new information from this region upon completion of the maps. The database will continue to be periodically updated as additional offshore mapping projects are finished along the Atlantic coast of Florida, the Gulf coast, the Pacific coast, and Alaska. The database will eventually contain the wind resource for all 50 states.

Incorporation of environmental exclusions and ocean-use factors impacting offshore wind development will be included in future editions of the database. These exclusions could include many factors including shipping lanes, marine habitat areas, submerged obstacles, military areas, and ocean-bottom topography. In addition, the database may be expanded to include other important characteristics such as wave power density, extreme wind and wave, ocean currents, and a number of other parameters important to the design of offshore wind turbines. Wave power resource estimates are the first type of water power data likely to be included in future versions of the database.

Acknowledgements

Thanks to George Scott and Dennis Elliott of the NREL wind resource assessment group for their work in processing and analyzing the offshore wind measurement data. We also thank the staff of AWS Truepower for producing the model data used to create the updated offshore wind resource maps. This paper was written at NREL in support of DOE under contract number DE-AC36-08-GO28308.

References

Applied Technology and Management Incorporated. (2007). RIWINDS Final Report Phase 1-Wind Energy Siting Study. 124pp.

Dhanju, A.; P. Whitaker, W Kempton. (2008). Assessing Offshore Wind Resources: An Accessible Methodology. Renewable Energy vol. 33(1) p. 55-64.

Elliott, D.; Holliday, C.; Barchet, W.; Foote, H.; Sandusky, W. (1987). Wind Energy Resource Atlas of the United States. DOE/CH 10093-4, Golden, Colorado: Solar Energy Research Institute.

Elliott, D.; Schwartz, M. (2006). Wind Resource Mapping for United States Offshore Areas: NREL/CP-500-40045. Golden, Colorado: NREL, 12 pp

Federal Register / Vol. 71, No. 1 / Tuesday, January 3, 2006 / Notices, page 127-131

Musial, W; Butterfield, C. (2004). Future for Offshore Wind Energy in the United States. NREL/CP-500-36313 in EnergyOcean Proceedings. Palm Beach, Florida

Musial, W. D. (2007). Offshore Wind: Viable Option for Coastal Regions of the United States. Marine Technology Society Journal, Published Fall 2007.

Thormahlen, L. (1999). Boundary Development of the Outer Continental Shelf. OCS Report MMS 99-006.: Department of the Interior Minerals Management Service. Lakewood, Colorado

U.S. Department of Energy, Energy Efficiency and Renewable Energy. (2008). 20% Wind Energy by 2030, Increasing Wind Energy's Contribution to U.S. Electricity Supply. Executive Summary. DOE/GO 102008-2578. Washington, D.C. 23 pp

U.S. Department of the Interior Minerals Management Service. (2007). Programmatic Environmental Impact Statement for Alternative Energy Development and Production and Alternate Use of Facilities on the Outer Continental Shelf – Draft Environmental Impact Statement.

U.S. Department of the Interior Minerals Management Service Web portal http://www.mms.gov/ld/Maps.htm.

U.S. Supreme Court, STATE OF NEW JERSEY, Plaintiff, v. STATE OF DELAWARE, Defendant. No. 134 Original, October Term 2007.

Appendix A. Data Sources

A1. State-to-State Boundary Data

Table A1. State-to-State Boundary Data

State	Source
California [1]	NREL digitized line from the MMS SLA line to the California/Oregon state line
Connecticut	http://www.ct.gov/dep/site/default.asp - http://www.nyswaterfronts.com/index.asp
Georgia	http://gis.state.ga.us/
Illinois	http://www.isgs.uiuc.edu/nsdihome/ - http://www.mcgi.state.mi.us/mgdl/
Louisiana [1]	http://atlas.lsu.edu/ - http://www.glo.state.tx.us/ - NREL digitized line from the MMS SLA line to the Texas/Louisiana state line
Maine	http://megis.maine.gov/
Maryland	http://www.marylandgis.net/SHAdata/
Massachusetts	http://www.whoi.edu/
Michigan	http://www.mcgi.state.mi.us/mgdl/
Minnesota	http://deli.dnr.state.mn.us/data_catalog.html
Mississippi [1]	http://www.maris.state.ms.us/ - http://atlas.lsu.edu/ - NREL digitized line from the MMS SLA line to the Mississippi/Alabama state line
New Jersey	http://www.state.nj.us/dep/njgs/ - http://www.nyswaterfronts.com/index.asp
New York	http://www.nyswaterfronts.com/index.asp
North Carolina	http://www.cgia.state.nc.us/ - http://www.ors.state.sc.us/digital/gisdata.asp

Ohio	http://www.dnr.state.oh.us/gims/ - http://www.mcgi.state.mi.us/mgdl/
Oregon [1]	NREL digitized line from the MMS SLA line to the Oregon/California and Oregon/Washington state lines
Pennsylvania	http://nationalatlas.gov/atlasftp.html - http://www.dnr.state.oh.us/gims/ - http://www.nyswaterfronts.com/index.asp
South Carolina	http://www.ors.state.sc.us/digital/gisdata.asp - http://gis.state.ga.us/
Texas [1]	http://www.glo.state.tx.us - NREL digitized line from the MMS SLA line to the Texas/Louisiana state line
Washington	http://www.ecy.wa.gov/services/gis/data/data.htm
[1] Environmental Systems Research Institute, Inc. Data & Maps 9.1 Detailed States	

A2. Wind Map Data

Table A2. Wind Map Datasets

State	Method	Preliminary Date	Final Date
California	Old Offshore Map	2002	2002
Connecticut	Old Offshore Map		2002
Delaware	Old Offshore Map	2002	2003
Georgia	Updated Offshore Map		2006
Hawaii	Old Offshore Map		2004
Illinois	Updated Offshore Map		2008
Indiana	Updated Offshore Map		2008
Louisiana	Updated Offshore Map		2007
Maine	Updated Offshore Map		2008
Maryland	Old Offshore Map	2002	2003
Massachusetts	Updated Offshore Map		2008
Michigan	Updated Offshore Map		2008
Minnesota	Updated Offshore Map		2008
New Hampshire	Updated Offshore Map		2008
New Jersey	Old Offshore Map	2002	2003
New York (Atlantic)	Old Offshore Map		2003
New York (Great Lakes)	Updated Offshore Map		2008
North Carolina	Old Offshore Map	2002	2003
Ohio	Updated Offshore Map		2008
Oregon	Old Offshore Map	2001	2002
Pennsylvania	Updated Offshore Map		2008
Rhode Island	Old Offshore Map		2002
South Carolina	Old Offshore Map		2006
Texas	Updated Offshore Map		2007
Virginia	Old Offshore Map	2002	2003
Washington	Old Offshore Map	2001	2002
Wisconsin	Updated Offshore Map		2008

Appendix B. Detailed maps and tables

B1. United States
B2. California
B3. Connecticut
B4. Delaware
B5. Georgia
B6. Hawaii
B7. Illinois
B8. Indiana
B9. Louisiana
B10. Maine
B11. Maryland
B12. Massachusetts
B13. Michigan
B14. Minnesota

B15. New Hampshire
B16. New Jersey
B17. New York
B18. North Carolina
B19. Ohio
B20. Oregon
B21. Pennsylvania
B22. Rhode Island
B23. South Carolina
B24. Texas
B25. Virginia
B26. Washington
B27. Wisconsin

NOTE: AWS Truepower produced the offshore wind resource estimates under the name AWS Truewind.

Appendix B. U.S. Detailed Wind Database Information by Element

Table B1. U.S. offshore wind resource by state and wind speed interval. Resource areas limited to >7.0 m/s at 90-m height and within 50 nm of shore.

State	Wind Speed at 90 m (m/s)							Total >7.0
	7.0 - 7.5 Area km² (MW)	7.5 - 8.0 Area km² (MW)	8.0 - 8.5 Area km² (MW)	8.5 - 9.0 Area km² (MW)	9.0 - 9.5 Area km² (MW)	9.5 - 10.0 Area km² (MW)	>10.0 Area km² (MW)	Area km² (MW)
California	11,439 (57,195)	24,864 (124,318)	23,059 (115,296)	22,852 (114,258)	13,185 (65,924)	15,231 (76,153)	6,926 (34,629)	117,555 (587,773)
Connecticut	530 (2,652)	702 (3,508)	40 (201)	0 (0)	0 (0)	0 (0)	0 (0)	1,272 (6,360)
Delaware	223 (1,116)	724 (3,618)	1,062 (5,310)	931 (4,657)	0 (0)	0 (0)	0 (0)	2,940 (14,701)
Georgia	3,820 (19,102)	7,741 (38,706)	523 (2,617)	0 (0)	0 (0)	0 (0)	0 (0)	12,085 (60,425)
Hawaii	18,873 (94,363)	42,298 (211,492)	33,042 (165,210)	13,913 (69,567)	7,779 (38,894)	6,720 (33,598)	4,852 (24,260)	127,477 (637,383)
Illinois	92 (458)	166 (830)	3,844 (19,222)	90 (450)	0 (0)	0 (0)	0 (0)	4,192 (20,960)
Indiana	82 (410)	216 (1,082)	286 (1,428)	0 (0)	0 (0)	0 (0)	0 (0)	584 (2,919)
Louisiana	48,043 (240,214)	15,032 (75,162)	0 (0)	0 (0)	0 (0)	0 (0)	0 (0)	63,075 (315,376)
Maine	906 (4,531)	1,142 (5,711)	1,976 (9,878)	3,331 (16,655)	8,429 (42,146)	15,485 (77,424)	42 (208)	31,311 (156,553)
Maryland	2,192 (10,960)	1,946 (9,732)	1,540 (7,700)	5,078 (25,390)	0 (0)	0 (0)	0 (0)	10,756 (53,782)
Massachusetts	202 (1,008)	526 (2,631)	1,639 (8,195)	3,606 (18,030)	20,351 (101,753)	13,674 (68,370)	0 (0)	39,997 (199,987)
Michigan	4,459 (22,297)	18,074 (90,368)	31,086 (155,430)	34,305 (171,524)	8,719 (43,593)	0 (0)	0 (0)	96,642 (483,212)

Table B1 (continued). U.S. offshore wind resource by state and wind speed interval. Resource areas limited to >7.0 m/s at 90-m height and within 50 nm of shore.

State	Wind Speed at 90 m (m/s)							
	7.0 - 7.5 Area km² (MW)	7.5 - 8.0 Area km² (MW)	8.0 - 8.5 Area km² (MW)	8.5 - 9.0 Area km² (MW)	9.0 - 9.5 Area km² (MW)	9.5 - 10.0 Area km² (MW)	>10.0 Area km² (MW)	Total >7.0 Area km² (MW)
Minnesota	3,102 (15,510)	994 (4,972)	0 (0)	0 (0)	0 (0)	0 (0)	0 (0)	4,096 (20,481)
New Hampshire	19 (93)	46 (229)	171 (853)	336 (1,678)	102 (508)	0 (0)	0 (0)	672 (3,361)
New Jersey	528 (2,642)	1,508 (7,538)	4,965 (24,827)	12,934 (64,668)	0 (0)	0 (0)	0 (0)	19,935 (99,675)
New York	1,105 (5,524)	4,358 (21,789)	8,324 (41,621)	2,876 (14,382)	7,453 (37,267)	5,322 (26,612)	0 (0)	29,439 (147,194)
North Carolina	1,847 (9,237)	4,098 (20,491)	13,655 (68,274)	39,875 (199,374)	16 (80)	0 (0)	0 (0)	59,491 (297,456)
Ohio	341 (1,705)	3,067 (15,333)	5,829 (29,147)	0 (0)	0 (0)	0 (0)	0 (0)	9,237 (46,186)
Oregon	388 (1,938)	1,493 (7,465)	8,644 (43,222)	13,925 (69,623)	7,394 (36,969)	6,065 (30,327)	5,986 (29,928)	43,894 (219,471)
Pennsylvania	34 (171)	211 (1,054)	1,679 (8,395)	0 (0)	0 (0)	0 (0)	0 (0)	1,924 (9,620)
Rhode Island	224 (1,121)	126 (630)	283 (1,417)	671 (3,356)	1,461 (7,307)	2,360 (11,799)	0 (0)	5,126 (25,631)
South Carolina	1,457 (7,283)	8,202 (41,010)	10,384 (51,919)	6,007 (30,033)	0 (0)	0 (0)	0 (0)	26,049 (130,244)
Texas[1]	2,019 (10,095)	24,823 (124,114)	16,556 (82,782)	12,273 (61,363)	0 (0)	0 (0)	0 (0)	55,671 (278,353)
Virginia	889 (4,446)	3,658 (18,289)	6,549 (32,743)	7,794 (38,970)	0 (0)	0 (0)	0 (0)	18,890 (94,448)

Table B1 (continued). U.S. offshore wind resource by state and wind speed interval. Resource areas limited to >7.0 m/s at 90-m height and within 50 nm of shore.

	Wind Speed at 90 m (m/s)						
	7.0 - 7.5	7.5 - 8.0	8.0 - 8.5	8.5 - 9.0	9.0 - 9.5	9.5 - 10.0	>10.0	Total >7.0
State	Area km^2 (MW)	Area km^2 (MW)	Area km^2 (MW)	Area km^2 (MW)	Area km^2 (MW)	Area km^2 (MW)	Area km^2 (MW)	Area km^2 (MW)
Washington	1,573 (7,863)	4,621 (23,107)	18,261 (91,304)	0 (0)	0 (0)	0 (0)	0 (0)	24,455 (122,274)
Wisconsin	3,715 (18,577)	3,405 (17,023)	7,761 (38,806)	8,417 (42,087)	0 (0)	0 (0)	0 (0)	23,298 (116,492)
Total	108,102 (540,510)	174,040 (870,202)	201,159 (1,005,795)	189,213 (946,063)	74,888 (374,441)	64,856 (324,282)	17,805 (89,024)	830,064 (4,150,319)

31

Table B1.1. U.S. offshore wind resource by state and water depth category. Resource areas limited to >7.0 m/s at 90-m height and within 50 nm of shore.

State	Shallow (0 - 30m) Area km^2 (MW)	Transitional (30 - 60m) Area km^2 (MW)	Deep (> 60m) Area km^2 (MW)	Total Area km^2 (MW)
California	871 (4,354)	2,091 (10,453)	114,593 (572,966)	117,555 (587,773)
Connecticut	1,152 (5,760)	118 (592)	2 (8)	1,272 (6,360)
Delaware	2,233 (11,164)	707 (3,537)	0 (0)	2,940 (14,701)
Georgia	9,643 (48,217)	2,442 (12,208)	0 (0)	12,085 (60,425)
Hawaii	460 (2,302)	1,106 (5,529)	125,910 (629,552)	127,477 (637,383)
Illinois	1,351 (6,757)	1,017 (5,086)	1,824 (9,118)	4,192 (20,960)
Indiana	584 (2,919)	0 (0)	0 (0)	584 (2,919)
Louisiana	36,698 (183,492)	8,380 (41,899)	17,997 (89,985)	63,075 (315,376)
Maine	3,145 (15,727)	3,048 (15,239)	25,117 (125,587)	31,311 (156,553)
Maryland	6,162 (30,809)	3,258 (16,292)	1,336 (6,681)	10,756 (53,782)
Massachusetts	9,570 (47,849)	11,612 (58,059)	18,816 (94,079)	39,997 (199,987)
Michigan	17,177 (85,887)	15,023 (75,115)	64,442 (322,210)	96,642 (483,212)
Minnesota	5 (25)	75 (376)	4,016 (20,080)	4,096 (20,481)
New Hampshire	116 (579)	171 (854)	386 (1,928)	672 (3,361)
New Jersey	8,696 (43,480)	10,910 (54,549)	329 (1,647)	19,935 (99,675)
New York	7,496 (37,481)	11,253 (56,263)	10,690 (53,450)	29,439 (147,194)
North Carolina	28,006 (140,029)	15,544 (77,719)	15,942 (79,708)	59,491 (297,456)
Ohio	9,237 (46,186)	0 (0)	0 (0)	9,237 (46,186)
Oregon	1,319 (6,595)	1,946 (9,728)	40,630 (203,148)	43,894 (219,471)

Table B1.1 (continued). U.S. offshore wind resource by state and water depth category. Resource areas limited to >7.0 m/s at 90-m height and within 50 nm of shore.

State	Shallow (0 - 30m) Area km^2 (MW)	Transitional (30 - 60m) Area km^2 (MW)	Deep (> 60m) Area km^2 (MW)	Total Area km^2 (MW)
Pennsylvania	1,451 (7,253)	473 (2,367)	0 (0)	1,924 (9,620)
Rhode Island	1,048 (5,238)	2,691 (13,453)	1,388 (6,940)	5,126 (25,631)
South Carolina	17,182 (85,911)	7,322 (36,611)	1,544 (7,722)	26,049 (130,244)
Texas[1]	31,370 (156,852)	15,637 (78,184)	8,663 (43,317)	55,671 (278,353)
Virginia	13,063 (65,317)	4,926 (24,631)	900 (4,500)	18,890 (94,448)
Washington	1,702 (8,509)	2,323 (11,614)	20,430 (102,151)	24,455 (122,274)
Wisconsin	4,502 (22,508)	3,529 (17,647)	15,267 (76,337)	23,298 (116,492)
Total	214,240 (1,071,202)	125,601 (628,004)	490,223 (2,451,113)	830,064 (4,150,319)

Table B1.2. U.S. offshore wind resource by state and distance from shore. Resource areas limited to >7.0 m/s at 90-m height and within 50 nm of shore.

State	0 - 3 nm Area km^2 (MW)	3 - 12 nm Area km^2 (MW)	12 - 50 nm Area km^2 (MW)	Total Area km^2 (MW)
California	2,447 (12,237)	20,824 (104,121)	94,283 (471,416)	117,555 (587,773)
Connecticut	1,272 (6,360)	0 (0)	0 (0)	1,272 (6,360)
Delaware	1,088 (5,438)	672 (3,359)	1,181 (5,904)	2,940 (14,701)
Georgia	633 (3,164)	2,692 (13,459)	8,760 (43,802)	12,085 (60,425)
Hawaii	13,455 (67,275)	19,794 (98,971)	94,228 (471,138)	127,477 (637,383)
Illinois	501 (2,507)	1,573 (7,867)	2,117 (10,586)	4,192 (20,960)
Indiana	337 (1,687)	247 (1,233)	0 (0)	584 (2,919)
Louisiana	7,944 (39,718)	11,554 (57,768)	43,578 (217,891)	63,075 (315,376)
Maine	5,483 (27,416)	7,919 (39,594)	17,909 (89,543)	31,311 (156,553)
Maryland	4,292 (21,459)	934 (4,670)	5,531 (27,653)	10,756 (53,782)
Massachusetts	5,392 (26,958)	7,087 (35,436)	27,519 (137,593)	39,997 (199,987)
Michigan	15,814 (79,068)	36,855 (184,277)	43,973 (219,867)	96,642 (483,212)
Minnesota	334 (1,670)	1,810 (9,052)	1,952 (9,759)	4,096 (20,481)
New Hampshire	154 (769)	364 (1,822)	154 (770)	672 (3,361)
New Jersey	2,271 (11,353)	3,665 (18,327)	13,999 (69,995)	19,935 (99,675)
New York	6,232 (31,159)	8,920 (44,602)	14,287 (71,434)	29,439 (147,194)
North Carolina	8,875 (44,373)	8,985 (44,923)	41,632 (208,160)	59,491 (297,456)
Ohio	2,013 (10,065)	4,153 (20,765)	3,071 (15,356)	9,237 (46,186)
Oregon	2,255 (11,274)	8,575 (42,876)	33,064 (165,320)	43,894 (219,471)

Table B1.2 (continued). U.S. offshore wind resource by state and distance from shore. Resource areas limited to >7.0 m/s at 90-m height and within 50 nm of shore.

State	0 - 3 nm Area km² (MW)	3 - 12 nm Area km² (MW)	12 - 50 nm Area km² (MW)	Total Area km² (MW)
Pennsylvania	425 (2,123)	1,206 (6,032)	293 (1,465)	1,924 (9,620)
Rhode Island	815 (4,074)	1,526 (7,630)	2,785 (13,927)	5,126 (25,631)
South Carolina	1,465 (7,323)	5,272 (26,358)	19,313 (96,563)	26,049 (130,244)
Texas[1]	21,285 (106,423)	3,128 (15,642)	31,257 (156,287)	55,671 (278,353)
Virginia	5,648 (28,239)	3,047 (15,233)	10,195 (50,976)	18,890 (94,448)
Washington	1,246 (6,230)	4,641 (23,205)	18,568 (92,839)	24,455 (122,274)
Wisconsin	4,381 (21,905)	9,452 (47,258)	9,466 (47,328)	23,298 (116,492)
Total	116,053 (580,267)	174,896 (874,481)	539,114 (2,695,572)	830,064 (4,150,319)

[1] Federal waters begin at 3 nm with the exception of Texas, which begins at 9 nm. For Texas, the area reported is for 0 - 9 nm; 9 - 12 nm; and 12 - 50 nm respectively.

Table B1.3. Offshore wind resource by state, wind speed interval, water depth and distance from shore within 50 nm of shore.

State	Wind Speed at 90m m/s	Distance from Shoreline									Total	
		0 - 3 nm[1]			3 - 12 nm			12 - 50 nm				
		Depth Category (m)			Depth Category (m)			Depth Category (m)				
		0 - 30	30 - 60	> 60	0 - 30	30 - 60	> 60	0 - 30	30 - 60	> 60		
		Area (km²) (MW)	Area (km²) (MW)	Area (km²) (MW)	Area (km²) (MW)	Area (km²) (MW)	Area (km²) (MW)	Area (km²) (MW)	Area (km²) (MW)	Area (km²) (MW)	Area (km²) (MW)	
California	7.0-7.5	266.2 (1,331)	236.2 (1,181)	257.4 (1,287)	100.9 (504)	456.8 (2,284)	4,554.0 (22,770)	7.7 (38)	22.9 (115)	5,536.8 (27,684)	11,438.9 (57,195)	
	7.5-8.0	239.1 (1,196)	256.9 (1,285)	189.6 (948)	78.8 (394)	595.7 (2,978)	3,854.6 (19,273)	0.0 (0)	32.9 (165)	19,616.1 (98,080)	24,863.7 (124,318)	
	8.0-8.5	125.2 (626)	178.2 (891)	281.8 (1,409)	7.1 (36)	105.8 (529)	4,539.1 (22,695)	0.0 (0)	0.0 (0)	17,822.2 (89,111)	23,059.3 (115,296)	
	8.5-9.0	43.2 (216)	141.7 (708)	176.4 (882)	0.6 (3)	38.0 (190)	4,559.8 (22,799)	0.0 (0)	0.0 (0)	17,892.0 (89,460)	22,851.6 (114,258)	
	9.0-9.5	2.1 (10)	18.8 (94)	14.9 (74)	0.0 (0)	0.9 (4)	988.0 (4,940)	0.0 (0)	0.0 (0)	12,160.2 (60,801)	13,184.8 (65,924)	
	9.5-10.0	0.0 (0)	6.0 (30)	13.9 (69)	0.0 (0)	0.0 (0)	656.1 (3,280)	0.0 (0)	0.0 (0)	14,554.8 (72,774)	15,230.7 (76,153)	
	>10.0	0.0 (0)	0.0 (0)	0.1 (1)	0.0 (0)	0.0 (0)	288.1 (1,441)	0.0 (0)	0.0 (0)	6,637.6 (33,188)	6,925.8 (34,629)	
Connecticut	7.0-7.5	500.1 (2,501)	30.2 (151)	0.0 (0)	0.0 (0)	0.0 (0)	0.0 (0)	0.0 (0)	0.0 (0)	0.0 (0)	530.4 (2,652)	
	7.5-8.0	617.3 (3,087)	83.0 (415)	1.2 (6)	0.0 (0)	0.0 (0)	0.0 (0)	0.0 (0)	0.0 (0)	0.0 (0)	701.6 (3,508)	
	8.0-8.5	34.6 (173)	5.0 (25)	0.4 (2)	0.0 (0)	0.0 (0)	0.0 (0)	0.0 (0)	0.0 (0)	0.0 (0)	40.1 (201)	
Delaware	7.0-7.5	223.2 (1,116)	0.0 (0)	0.0 (0)	0.0 (0)	0.0 (0)	0.0 (0)	0.0 (0)	0.0 (0)	0.0 (0)	223.2 (1,116)	
	7.5-8.0	716.5 (3,583)	2.0 (10)	0.0 (0)	5.2 (26)	0.0 (0)	0.0 (0)	0.0 (0)	0.0 (0)	0.0 (0)	723.7 (3,618)	
	8.0-8.5	135.4 (677)	10.5 (53)	0.0 (0)	658.0 (3,290)	8.6 (43)	0.0 (0)	240.4 (1,202)	8.9 (44)	0.0 (0)	1,061.9 (5,310)	
	8.5-9.0	0.0 (0)	0.0 (0)	0.0 (0)	0.0 (0)	0.0 (0)	0.0 (0)	254.0 (1,270)	677.4 (3,387)	0.0 (0)	931.4 (4,657)	

Table B.1.3 (continued) Offshore wind resource by state, wind speed interval, water depth and distance from shore within 50 nm of shore.

State	Wind Speed at 90m m/s	Distance from Shoreline										
		0 - 3 nm[1]			3 - 12 nm			12 - 50 nm			Total
		Depth Category (m)			Depth Category (m)			Depth Category (m)			
		0 - 30 Area (km²) (MW)	30 - 60 Area (km²) (MW)	> 60 Area (km²) (MW)	0 - 30 Area (km²) (MW)	30 - 60 Area (km²) (MW)	> 60 Area (km²) (MW)	0 - 30 Area (km²) (MW)	30 - 60 Area (km²) (MW)	> 60 Area (km²) (MW)	Area (km²) (MW)
Georgia	7.0-7.5	547.5 (2,737)	0.0 (0)	0.0 (0)	2,162.3 (10,811)	0.0 (0)	0.0 (0)	1,110.6 (5,553)	0.0 (0)	0.0 (0)	3,820.4 (19,102)
	7.5-8.0	85.2 (426)	0.0 (0)	0.0 (0)	529.6 (2,648)	0.0 (0)	0.0 (0)	5,204.3 (26,021)	1,922.1 (9,610)	0.0 (0)	7,741.2 (38,706)
	8.0-8.5	0.0 (0)	0.0 (0)	0.0 (0)	0.0 (0)	0.0 (0)	0.0 (0)	3.9 (19)	519.5 (2,598)	0.0 (0)	523.4 (2,617)
Hawaii	7.0-7.5	111.5 (557)	97.2 (486)	2,631.5 (13,157)	0.0 (0)	42.6 (213)	2,213.3 (11,067)	44.4 (222)	116.8 (584)	13,615.3 (68,077)	18,872.6 (94,363)
	7.5-8.0	65.5 (328)	107.0 (535)	2,404.0 (12,020)	0.0 (0)	145.0 (725)	5,051.6 (25,258)	6.9 (34)	250.7 (1,254)	34,267.6 (171,338)	42,298.3 (211,492)
	8.0-8.5	92.2 (461)	114.8 (574)	2,364.3 (11,822)	0.0 (0)	15.0 (75)	4,755.7 (23,778)	0.2 (1)	0.3 (1)	25,699.5 (128,498)	33,042.0 (165,210)
	8.5-9.0	65.6 (328)	67.0 (335)	2,105.2 (10,526)	0.0 (0)	0.0 (0)	2,625.5 (13,128)	0.0 (0)	0.0 (0)	9,050.0 (45,250)	13,913.3 (69,567)
	9.0-9.5	25.7 (129)	39.0 (195)	995.6 (4,978)	0.0 (0)	0.0 (0)	1,853.0 (9,265)	0.0 (0)	0.0 (0)	4,865.5 (24,328)	7,778.8 (38,894)
	9.5-10.0	22.0 (110)	39.8 (199)	666.0 (3,330)	0.0 (0)	0.0 (0)	1,064.9 (5,324)	0.0 (0)	0.0 (0)	4,926.8 (24,634)	6,719.6 (33,598)
	>10.0	26.3 (132)	70.5 (353)	1,344.0 (6,720)	0.0 (0)	0.0 (0)	2,027.6 (10,138)	0.0 (0)	0.0 (0)	1,383.5 (6,918)	4,852.0 (24,260)
Illinois	7.0-7.5	91.6 (458)	0.0 (0)	0.0 (0)	0.0 (0)	0.0 (0)	0.0 (0)	0.0 (0)	0.0 (0)	0.0 (0)	91.6 (458)
	7.5-8.0	164.6 (823)	0.0 (0)	0.0 (0)	0.0 (0)	0.1 (1)	0.0 (0)	0.0 (0)	1.3 (6)	0.0 (0)	166.0 (830)
	8.0-8.5	244.2 (1,221)	1.0 (5)	0.0 (0)	830.4 (4,152)	512.2 (2,561)	140.7 (704)	7.1 (35)	426.1 (2,130)	1,682.8 (8,414)	3,844.5 (19,222)
	8.5-9.0	0.0 (0)	0.0 (0)	0.0 (0)	13.4 (67)	76.6 (383)	0.0 (0)	0.0 (0)	0.0 (0)	0.0 (0)	89.9 (450)

Table B.1.3 (continued) Offshore wind resource by state, wind speed interval, water depth and distance from shore within 50 nm of shore.

Distance from Shoreline — Depth Category (m). Each cell: Area (km²) with (MW) below in parentheses.

State	Wind Speed at 90m (m/s)	0 - 3 nm[1] 0-30	0 - 3 nm 30-60	0 - 3 nm >60	3 - 12 nm 0-30	3 - 12 nm 30-60	3 - 12 nm >60	12 - 50 nm 0-30	12 - 50 nm 30-60	12 - 50 nm >60	Total
Indiana	7.0-7.5	82.1 (410)	0.0 (0)	0.0 (0)	0.0 (0)	0.0 (0)	0.0 (0)	0.0 (0)	0.0 (0)	0.0 (0)	82.1 (410)
	7.5-8.0	153.8 (769)	0.0 (0)	0.0 (0)	62.6 (313)	0.0 (0)	0.0 (0)	0.0 (0)	0.0 (0)	0.0 (0)	216.3 (1,082)
	8.0-8.5	101.5 (507)	0.0 (0)	0.0 (0)	184.0 (920)	0.0 (0)	0.0 (0)	0.0 (0)	0.0 (0)	0.0 (0)	285.5 (1,428)
Louisiana	7.0-7.5	7,759.6 (38,798)	28.6 (143)	0.0 (0)	7,825.2 (39,126)	643.4 (3,217)	1,460.4 (7,302)	11,163.9 (55,819)	5,479.1 (27,396)	13,682.4 (68,412)	48,042.7 (240,214)
	7.5-8.0	155.3 (776)	0.0 (0)	0.0 (0)	1,624.5 (8,123)	0.0 (0)	0.0 (0)	8,169.9 (40,850)	2,228.7 (11,143)	2,854.1 (14,271)	15,032.5 (75,162)
Maine	7.0-7.5	787.0 (3,935)	91.2 (456)	11.9 (59)	7.8 (39)	4.8 (24)	3.5 (18)	0.0 (0)	0.0 (0)	0.0 (0)	906.2 (4,531)
	7.5-8.0	797.2 (3,986)	285.4 (1,427)	19.4 (97)	6.7 (33)	19.6 (98)	14.1 (70)	0.0 (0)	0.0 (0)	0.0 (0)	1,142.3 (5,711)
	8.0-8.5	777.0 (3,885)	440.8 (2,204)	74.2 (371)	63.4 (317)	385.6 (1,928)	234.5 (1,173)	0.0 (0)	0.0 (0)	0.0 (0)	1,975.6 (9,878)
	8.5-9.0	513.4 (2,567)	614.0 (3,070)	157.6 (788)	18.2 (91)	219.1 (1,095)	1,401.9 (7,010)	0.0 (0)	0.0 (0)	406.8 (2,034)	3,331.1 (16,655)
	9.0-9.5	142.2 (711)	390.0 (1,950)	309.2 (1,546)	25.9 (129)	469.0 (2,345)	3,504.1 (17,520)	0.0 (0)	57.8 (289)	3,530.9 (17,655)	8,429.2 (42,146)
	9.5-10.0	5.5 (28)	24.9 (124)	42.3 (211)	1.0 (5)	38.3 (191)	1,459.8 (7,299)	0.0 (0)	7.4 (37)	13,905.6 (69,528)	15,484.7 (77,424)
	>10.0	0.0 (0)	0.0 (0)	0.0 (0)	0.0 (0)	0.0 (0)	41.6 (208)	0.0 (0)	0.0 (0)	0.0 (0)	41.6 (208)
Maryland	7.0-7.5	2,175.4 (10,877)	16.6 (83)	0.0 (0)	0.0 (0)	0.0 (0)	0.0 (0)	0.0 (0)	0.0 (0)	0.0 (0)	2,192.0 (10,960)
	7.5-8.0	1,922.5 (9,613)	14.0 (70)	0.0 (0)	9.9 (49)	0.0 (0)	0.0 (0)	0.0 (0)	0.0 (0)	0.0 (0)	1,946.4 (9,732)

Table B.1.3 (continued) Offshore wind resource by state, wind speed interval, water depth and distance from shore within 50 nm of shore.

State	Wind Speed at 90m m/s	0 - 3 nm[1] 0-30 Area (km²) (MW)	0 - 3 nm 30-60 Area (km²) (MW)	0 - 3 nm >60 Area (km²) (MW)	3 - 12 nm 0-30 Area (km²) (MW)	3 - 12 nm 30-60 Area (km²) (MW)	3 - 12 nm >60 Area (km²) (MW)	12 - 50 nm 0-30 Area (km²) (MW)	12 - 50 nm 30-60 Area (km²) (MW)	12 - 50 nm >60 Area (km²) (MW)	Total Area (km²) (MW)
Maryland (cont.)	8.0-8.5	163.3 (817)	0.0 (0)	0.0 (0)	924.0 (4,620)	0.0 (0)	0.0 (0)	435.4 (2,177)	17.2 (86)	0.0 (0)	1,539.9 (7,700)
	8.5-9.0	0.0 (0)	0.0 (0)	0.0 (0)	0.0 (0)	0.0 (0)	0.0 (0)	531.3 (2,656)	3,210.7 (16,053)	1,336.1 (6,681)	5,078.1 (25,390)
Massachusetts	7.0-7.5	201.6 (1,008)	0.0 (0)	0.0 (0)	0.0 (0)	0.0 (0)	0.0 (0)	0.0 (0)	0.0 (0)	0.0 (0)	201.6 (1,008)
	7.5-8.0	521.4 (2,607)	4.7 (23)	0.0 (0)	0.0 (0)	0.0 (0)	0.0 (0)	0.0 (0)	0.0 (0)	0.0 (0)	526.1 (2,631)
	8.0-8.5	927.4 (4,637)	327.3 (1,636)	28.6 (143)	78.2 (391)	152.0 (760)	125.5 (628)	0.0 (0)	0.0 (0)	0.0 (0)	1,639.1 (8,195)
	8.5-9.0	1,508.2 (7,541)	378.1 (1,890)	12.6 (63)	315.0 (1,575)	354.5 (1,773)	812.2 (4,061)	11.4 (57)	23.5 (118)	190.4 (952)	3,606.0 (18,030)
	9.0-9.5	1,137.0 (5,685)	322.6 (1,613)	20.0 (100)	2,696.9 (13,484)	1,418.6 (7,093)	1,006.5 (5,033)	1,689.9 (8,449)	5,051.8 (25,259)	7,007.4 (35,037)	20,350.7 (101,753)
	9.5-10.0	2.0 (10)	0.0 (0)	0.0 (0)	8.6 (43)	119.2 (596)	0.0 (0)	472.1 (2,361)	3,459.5 (17,298)	9,612.5 (48,063)	13,674.0 (68,370)
Michigan	7.0-7.5	3,411.4 (17,057)	330.6 (1,653)	243.7 (1,218)	142.6 (713)	49.8 (249)	268.0 (1,340)	1.2 (6)	0.0 (0)	12.1 (61)	4,459.3 (22,297)
	7.5-8.0	4,632.7 (23,163)	1,258.2 (6,291)	635.3 (3,177)	3,069.0 (15,345)	2,916.4 (14,582)	2,991.0 (14,955)	14.2 (71)	108.5 (542)	2,448.4 (12,242)	18,073.7 (90,368)
	8.0-8.5	2,588.1 (12,940)	1,253.3 (6,267)	769.0 (3,845)	2,770.6 (13,853)	5,889.7 (29,448)	9,411.9 (47,059)	30.9 (155)	876.1 (4,381)	7,496.4 (37,482)	31,085.9 (155,430)
	8.5-9.0	138.7 (693)	156.2 (781)	396.5 (1,983)	145.1 (726)	550.6 (2,753)	8,580.8 (42,904)	118.5 (592)	1,383.9 (6,920)	22,834.5 (114,172)	34,304.7 (171,524)
	9.0-9.5	0.0 (0)	0.0 (0)	0.0 (0)	0.0 (0)	0.0 (0)	69.9 (350)	114.6 (573)	249.6 (1,248)	8,284.5 (41,423)	8,718.7 (43,593)
Minnesota	7.0-7.5	5.1 (25)	43.6 (218)	272.6 (1,363)	0.0 (0)	31.5 (158)	1,718.1 (8,590)	0.0 (0)	0.0 (0)	1,031.1 (5,155)	3,101.9 (15,510)
	7.5-8.0	0.0 (0)	0.1 (0)	12.7 (64)	0.0 (0)	0.0 (0)	60.8 (304)	0.0 (0)	0.0 (0)	920.7 (4,603)	994.3 (4,972)

Table B.1.3 (continued) Offshore wind resource by state, wind speed interval, water depth and distance from shore within 50 nm of shore.

State	Wind Speed at 90m m/s	Distance from Shoreline												
		0 - 3 nm[1]			3 - 12 nm			12 - 50 nm				Total
		Depth Category (m)			Depth Category (m)			Depth Category (m)				
		0 - 30 Area (km²) (MW)	30 - 60 Area (km²) (MW)	> 60 Area (km²) (MW)	0 - 30 Area (km²) (MW)	30 - 60 Area (km²) (MW)	> 60 Area (km²) (MW)	0 - 30 Area (km²) (MW)	30 - 60 Area (km²) (MW)	> 60 Area (km²) (MW)		Area (km²) (MW)
New Hampshire	7.0-7.5	18.6 (93)	0.0 (0)	0.0 (0)	0.0 (0)	0.0 (0)	0.0 (0)	0.0 (0)	0.0 (0)	0.0 (0)		18.6 (93)
	7.5-8.0	45.8 (229)	0.0 (0)	0.0 (0)	0.0 (0)	0.0 (0)	0.0 (0)	0.0 (0)	0.0 (0)	0.0 (0)		45.8 (229)
	8.0-8.5	44.6 (223)	29.7 (148)	0.0 (0)	6.9 (34)	75.5 (378)	14.0 (70)	0.0 (0)	0.0 (0)	0.0 (0)		170.6 (853)
	8.5-9.0	0.0 (0)	8.0 (40)	7.2 (36)	0.0 (0)	12.4 (62)	255.7 (1,279)	0.0 (0)	10.1 (51)	42.2 (211)		335.7 (1,678)
	9.0-9.5	0.0 (0)	0.0 (0)	0.0 (0)	0.0 (0)	0.0 (0)	0.0 (0)	0.0 (0)	35.2 (176)	66.4 (332)		101.6 (508)
New Jersey	7.0-7.5	520.2 (2,601)	0.0 (0)	0.0 (0)	8.2 (41)	0.0 (0)	0.0 (0)	0.0 (0)	0.0 (0)	0.0 (0)		528.5 (2,642)
	7.5-8.0	960.4 (4,802)	0.0 (0)	0.0 (0)	518.6 (2,593)	27.9 (140)	0.7 (4)	0.0 (0)	0.0 (0)	0.0 (0)		1,507.6 (7,538)
	8.0-8.5	776.9 (3,885)	0.0 (0)	0.0 (0)	2,764.8 (13,824)	127.1 (636)	3.2 (16)	899.4 (4,497)	372.4 (1,862)	21.6 (108)		4,965.4 (24,827)
	8.5-9.0	13.1 (66)	0.0 (0)	0.0 (0)	214.7 (1,074)	0.0 (0)	0.0 (0)	2,019.5 (10,098)	10,382.4 (51,912)	303.8 (1,519)		12,933.5 (64,668)
New York	7.0-7.5	1,047.9 (5,239)	54.2 (271)	0.0 (0)	2.8 (14)	0.0 (0)	0.0 (0)	0.0 (0)	0.0 (0)	0.0 (0)		1,104.8 (5,524)
	7.5-8.0	1,983.1 (9,915)	354.7 (1,774)	11.6 (58)	472.7 (2,363)	609.0 (3,045)	866.8 (4,334)	0.0 (0)	0.0 (0)	59.9 (300)		4,357.8 (21,789)
	8.0-8.5	1,329.3 (6,646)	602.9 (3,015)	196.5 (982)	375.6 (1,878)	535.7 (2,679)	3,122.0 (15,610)	42.2 (211)	12.6 (63)	2,107.4 (10,537)		8,324.1 (41,621)
	8.5-9.0	535.1 (2,675)	110.6 (553)	0.6 (3)	1,042.6 (5,213)	93.8 (469)	0.0 (0)	261.5 (1,307)	727.8 (3,639)	104.5 (522)		2,876.4 (14,382)
	9.0-9.5	5.4 (27)	0.0 (0)	0.0 (0)	391.9 (1,959)	1,407.6 (7,038)	0.0 (0)	6.4 (32)	4,926.1 (24,631)	716.0 (3,580)		7,453.3 (37,267)
	9.5-10.0	0.0 (0)	0.0 (0)	0.0 (0)	0.0 (0)	0.0 (0)	0.0 (0)	0.0 (0)	1,817.6 (9,088)	3,504.8 (17,524)		5,322.4 (26,612)

40

Table B.1.3 (continued) Offshore wind resource by state, wind speed interval, water depth and distance from shore within 50 nm of shore.

State	Wind Speed at 90m m/s	Distance from Shoreline									Total
		0 - 3 nm[1]			3 - 12 nm			12 - 50 nm			
		Depth Category (m)			Depth Category (m)			Depth Category (m)			
		0 - 30 Area (km²) (MW)	30 - 60 Area (km²) (MW)	> 60 Area (km²) (MW)	0 - 30 Area (km²) (MW)	30 - 60 Area (km²) (MW)	> 60 Area (km²) (MW)	0 - 30 Area (km²) (MW)	30 - 60 Area (km²) (MW)	> 60 Area (km²) (MW)	Area (km²) (MW)
North Carolina	7.0-7.5	1,847.4 (9,237)	0.0 (0)	0.0 (0)	0.0 (0)	0.0 (0)	0.0 (0)	0.0 (0)	0.0 (0)	0.0 (0)	1,847.4 (9,237)
	7.5-8.0	3,002.1 (15,010)	0.0 (0)	0.0 (0)	1,096.1 (5,481)	0.0 (0)	0.0 (0)	0.0 (0)	0.0 (0)	0.0 (0)	4,098.3 (20,491)
	8.0-8.5	3,920.0 (19,600)	0.1 (1)	0.1 (0)	4,653.8 (23,269)	0.0 (0)	0.0 (0)	4,034.9 (20,174)	746.5 (3,733)	299.2 (1,496)	13,654.7 (68,274)
	8.5-9.0	105.0 (525)	0.0 (0)	0.0 (0)	2,964.1 (14,821)	261.4 (1,307)	8.9 (45)	6,382.4 (31,912)	14,519.6 (72,598)	15,633.3 (78,167)	39,874.8 (199,374)
	9.0-9.5	0.0 (0)	0.0 (0)	0.0 (0)	0.0 (0)	0.0 (0)	0.0 (0)	0.0 (0)	16.0 (80)	0.0 (0)	16.0 (80)
Ohio	7.0-7.5	341.0 (1,705)	0.0 (0)	0.0 (0)	0.1 (0)	0.0 (0)	0.0 (0)	0.0 (0)	0.0 (0)	0.0 (0)	341.0 (1,705)
	7.5-8.0	1,106.8 (5,534)	0.0 (0)	0.0 (0)	1,959.8 (9,799)	0.0 (0)	0.0 (0)	0.0 (0)	0.0 (0)	0.0 (0)	3,066.7 (15,333)
	8.0-8.5	565.2 (2,826)	0.0 (0)	0.0 (0)	2,193.1 (10,965)	0.0 (0)	0.0 (0)	3,071.2 (15,356)	0.0 (0)	0.0 (0)	5,829.5 (29,147)
Oregon	7.0-7.5	355.8 (1,779)	20.6 (103)	0.0 (0)	0.9 (4)	9.2 (46)	1.2 (6)	0.0 (0)	0.0 (0)	0.0 (0)	387.6 (1,938)
	7.5-8.0	523.0 (2,615)	319.2 (1,596)	37.6 (188)	46.5 (232)	231.8 (1,159)	335.0 (1,675)	0.0 (0)	0.0 (0)	0.0 (0)	1,493.0 (7,465)
	8.0-8.5	198.3 (991)	277.0 (1,385)	6.6 (33)	19.0 (95)	595.7 (2,978)	2,558.5 (12,792)	0.0 (0)	0.0 (0)	4,989.5 (24,947)	8,644.4 (43,222)
	8.5-9.0	64.1 (320)	98.8 (494)	0.6 (3)	0.0 (0)	108.0 (540)	1,967.2 (9,836)	0.0 (0)	45.6 (228)	11,640.3 (58,201)	13,924.6 (69,623)
	9.0-9.5	64.1 (321)	55.4 (277)	38.6 (193)	0.0 (0)	32.6 (163)	614.9 (3,074)	0.0 (0)	0.0 (0)	6,588.2 (32,941)	7,393.7 (36,969)
	9.5-10.0	47.3 (237)	80.4 (402)	14.6 (73)	0.0 (0)	33.8 (169)	634.6 (3,173)	0.0 (0)	0.0 (0)	5,254.6 (26,273)	6,065.3 (30,327)
	>10.0	0.2 (1)	19.4 (97)	33.3 (166)	0.0 (0)	18.2 (91)	1,368.5 (6,843)	0.0 (0)	0.0 (0)	4,546.0 (22,730)	5,985.5 (29,928)

Table B.1.3 (continued) Offshore wind resource by state, wind speed interval, water depth and distance from shore within 50 nm of shore.

State	Wind Speed at 90m m/s	0 - 3 nm[1]						3 - 12 nm						12 - 50 nm						Total	
		0 - 30		30 - 60		> 60		0 - 30		30 - 60		> 60		0 - 30		30 - 60		> 60			
		Area (km^2)	(MW)	Area (km^2)	(MW)	Area (km^2)	(MW)	Area (km^2)	(MW)	Area (km^2)	(MW)	Area (km^2)	(MW)	Area (km^2)	(MW)	Area (km^2)	(MW)	Area (km^2)	(MW)	Area (km^2)	(MW)
Pennsylvania	7.0-7.5	34.2	(171)	0.0	(0)	0.0	(0)	0.0	(0)	0.0	(0)	0.0	(0)	0.0	(0)	0.0	(0)	0.0	(0)	34.2	(171)
	7.5-8.0	113.4	(567)	0.0	(0)	0.0	(0)	53.2	(266)	44.2	(221)	0.0	(0)	0.0	(0)	0.0	(0)	0.0	(0)	210.8	(1,054)
	8.0-8.5	276.2	(1,381)	0.9	(4)	0.0	(0)	750.8	(3,754)	358.1	(1,790)	0.0	(0)	222.9	(1,114)	70.2	(351)	0.0	(0)	1,679.1	(8,395)
Rhode Island	7.0-7.5	216.3	(1,082)	8.0	(40)	0.0	(0)	0.0	(0)	0.0	(0)	0.0	(0)	0.0	(0)	0.0	(0)	0.0	(0)	224.3	(1,121)
	7.5-8.0	123.2	(616)	2.8	(14)	0.0	(0)	0.0	(0)	0.0	(0)	0.0	(0)	0.0	(0)	0.0	(0)	0.0	(0)	126.0	(630)
	8.0-8.5	139.9	(700)	39.8	(199)	0.0	(0)	34.5	(173)	69.2	(346)	0.0	(0)	0.0	(0)	0.0	(0)	0.0	(0)	283.5	(1,417)
	8.5-9.0	120.3	(602)	93.0	(465)	0.0	(0)	183.6	(918)	274.2	(1,371)	0.0	(0)	0.0	(0)	0.0	(0)	0.0	(0)	671.2	(3,356)
	9.0-9.5	53.7	(269)	17.6	(88)	0.0	(0)	176.0	(880)	782.7	(3,914)	0.0	(0)	0.0	(0)	430.2	(2,151)	1.3	(6)	1,461.5	(7,307)
	9.5-10.0	0.0	(0)	0.0	(0)	0.0	(0)	0.0	(0)	5.8	(29)	0.0	(0)	0.0	(0)	967.2	(4,836)	1,386.8	(6,934)	2,359.8	(11,799)
South Carolina	7.0-7.5	848.2	(4,241)	0.0	(0)	0.0	(0)	608.4	(3,042)	0.0	(0)	0.0	(0)	0.0	(0)	0.0	(0)	0.0	(0)	1,456.5	(7,283)
	7.5-8.0	593.5	(2,968)	0.0	(0)	0.0	(0)	3,053.9	(15,269)	0.0	(0)	0.0	(0)	4,267.6	(21,338)	287.0	(1,435)	0.0	(0)	8,202.1	(41,010)
	8.0-8.5	22.9	(115)	0.0	(0)	0.0	(0)	1,609.4	(8,047)	0.0	(0)	0.0	(0)	4,151.4	(20,757)	3,925.6	(19,628)	674.5	(3,372)	10,383.7	(51,919)
	8.5-9.0	0.0	(0)	0.0	(0)	0.0	(0)	0.0	(0)	0.0	(0)	0.0	(0)	2,027.0	(10,135)	3,109.7	(15,548)	869.9	(4,349)	6,006.5	(30,033)

42

Table B.1.3 (cont'd) Offshore wind resource by state, wind speed interval, water depth and distance from shore within 50 nm.

State	Wind Speed at 90m m/s	Distance from Shoreline									Total
		0 - 3 nm[1]			3 - 12 nm			12 - 50 nm			
		Depth Category (m)			Depth Category (m)			Depth Category (m)			
		0 - 30	30 - 60	> 60	0 - 30	30 - 60	> 60	0 - 30	30 - 60	> 60	
		Area (km²) (MW)	Area (km²) (MW)	Area (km²) (MW)	Area (km²) (MW)	Area (km²) (MW)	Area (km²) (MW)	Area (km²) (MW)	Area (km²) (MW)	Area (km²) (MW)	Area (km²) (MW)
Texas[1]	7.0-7.5	1,785.5 (8,928)	0.0 (0)	0.1 (1)	96.2 (481)	0.0 (0)	0.0 (0)	137.2 (686)	0.0 (0)	0.0 (0)	2,019.1 (10,095)
	7.5-8.0	9,046.1 (45,230)	0.0 (0)	0.0 (0)	1,165.2 (5,826)	0.0 (0)	0.1 (0)	8,045.2 (40,226)	6,174.7 (30,873)	391.5 (1,958)	24,822.7 (124,114)
	8.0-8.5	5,928.5 (29,643)	125.4 (627)	0.0 (0)	742.1 (3,711)	136.9 (684)	0.0 (0)	583.5 (2,918)	4,442.8 (22,214)	4,597.1 (22,985)	16,556.3 (82,782)
	8.5-9.0	3,798.0 (18,990)	601.0 (3,005)	0.0 (0)	42.9 (215)	945.1 (4,726)	0.0 (0)	0.0 (0)	3,210.9 (16,055)	3,674.6 (18,373)	12,272.5 (61,363)
Virginia	7.0-7.5	889.2 (4,446)	0.0 (0)	0.0 (0)	0.0 (0)	0.0 (0)	0.0 (0)	0.0 (0)	0.0 (0)	0.0 (0)	889.2 (4,446)
	7.5-8.0	3,605.5 (18,028)	15.3 (77)	0.0 (0)	37.0 (185)	0.0 (0)	0.0 (0)	0.0 (0)	0.0 (0)	0.0 (0)	3,657.8 (18,289)
	8.0-8.5	1,135.8 (5,679)	2.0 (10)	0.0 (0)	2,985.9 (14,929)	0.0 (0)	0.0 (0)	2,355.9 (11,780)	69.1 (345)	0.0 (0)	6,548.7 (32,743)
	8.5-9.0	0.0 (0)	0.0 (0)	0.0 (0)	23.7 (119)	0.0 (0)	0.0 (0)	2,030.5 (10,152)	4,839.8 (24,199)	900.0 (4,500)	7,794.0 (38,970)
Washington	7.0-7.5	622.9 (3,115)	133.5 (667)	57.3 (286)	231.9 (1,159)	350.2 (1,751)	176.7 (884)	0.0 (0)	0.0 (0)	0.2 (1)	1,572.7 (7,863)
	7.5-8.0	370.6 (1,853)	0.0 (0)	0.0 (0)	203.0 (1,015)	841.6 (4,208)	1,173.0 (5,865)	0.0 (0)	0.0 (0)	2,033.1 (10,166)	4,621.4 (23,107)
	8.0-8.5	61.7 (308)	0.0 (0)	0.0 (0)	211.7 (1,059)	977.9 (4,889)	475.1 (2,375)	0.0 (0)	19.6 (98)	16,514.8 (82,574)	18,260.7 (91,304)
Wisconsin	7.0-7.5	1,053.8 (5,269)	268.2 (1,341)	145.8 (729)	349.9 (1,749)	616.6 (3,083)	1,085.0 (5,425)	0.0 (0)	34.2 (171)	162.0 (810)	3,715.4 (18,577)
	7.5-8.0	1,417.2 (7,086)	235.5 (1,177)	23.7 (119)	361.7 (1,808)	223.0 (1,115)	779.3 (3,896)	0.0 (0)	0.0 (0)	364.2 (1,821)	3,404.5 (17,023)
	8.0-8.5	830.6 (4,153)	372.7 (1,863)	0.0 (0)	451.4 (2,257)	1,386.1 (6,930)	2,539.0 (12,695)	0.0 (0)	21.3 (107)	2,160.1 (10,800)	7,761.1 (38,806)
	8.5-9.0	25.5 (128)	8.1 (41)	0.0 (0)	11.7 (58)	300.8 (1,504)	1,347.3 (6,737)	0.0 (0)	62.8 (314)	6,661.0 (33,305)	8,417.3 (42,087)

[1] Federal waters begin at 3 nm with the exception of Texas, which begins at 9 nm. For Texas, area reported is 0 - 9, 9 - 12; and 12 - 50 nm.

43

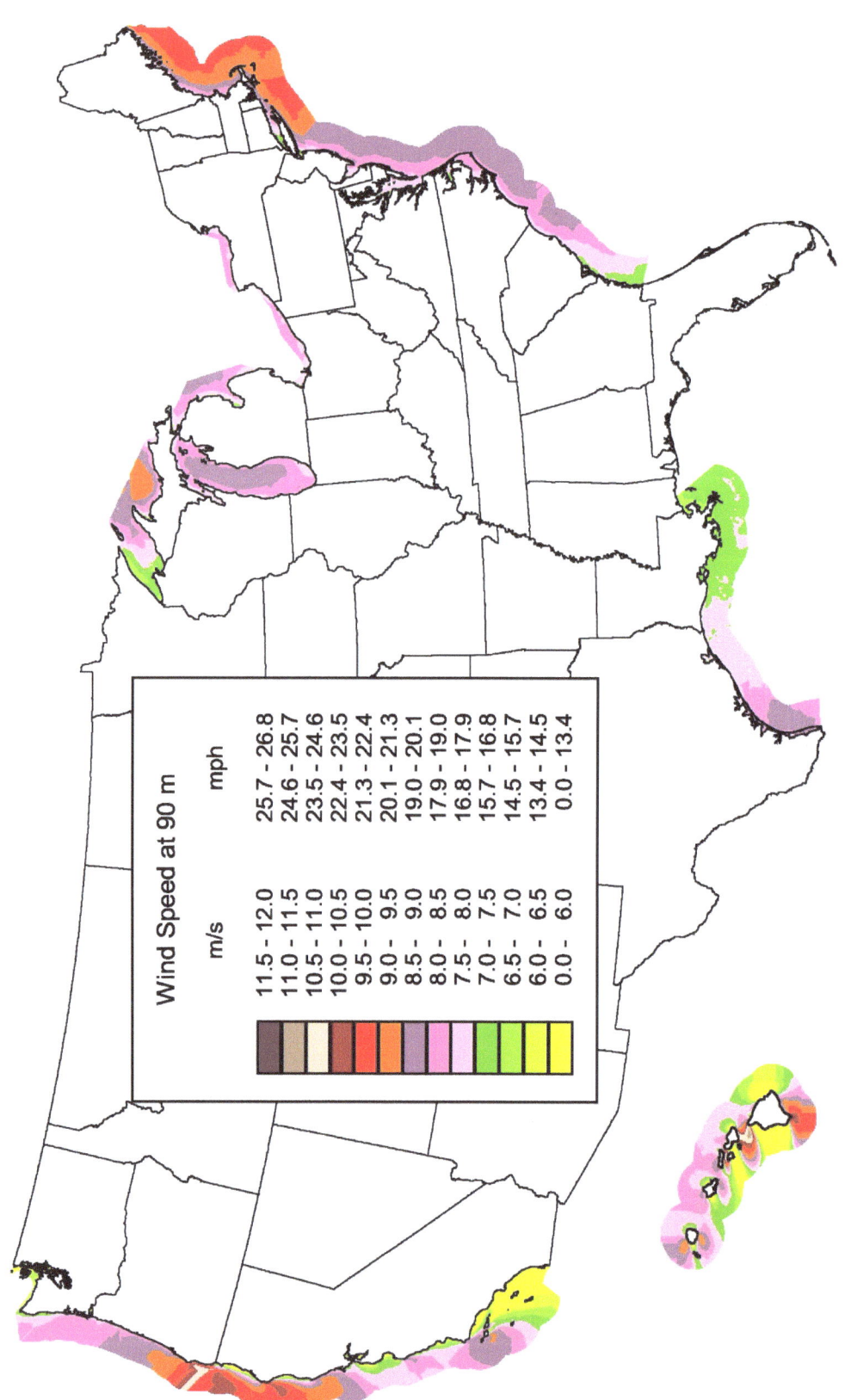

Wind Speed at 90 m

m/s	mph
11.5 - 12.0	25.7 - 26.8
11.0 - 11.5	24.6 - 25.7
10.5 - 11.0	23.5 - 24.6
10.0 - 10.5	22.4 - 23.5
9.5 - 10.0	21.3 - 22.4
9.0 - 9.5	20.1 - 21.3
8.5 - 9.0	19.0 - 20.1
8.0 - 8.5	17.9 - 19.0
7.5 - 8.0	16.8 - 17.9
7.0 - 7.5	15.7 - 16.8
6.5 - 7.0	14.5 - 15.7
6.0 - 6.5	13.4 - 14.5
0.0 - 6.0	0.0 - 13.4

Figure B1. United States offshore wind resource at 90 m above the surface.

44

Table B2. California offshore wind resource by wind speed interval, water depth and distance from shore within 50 nm of shore.

Depth Category	Distance from Shore (nm)								
	0 - 3			3 - 12			12 - 50		
90 m Wind Speed Interval (m/s)	Shallow (0 - 30 m) Area km² (MW)	Transitional (30 - 60m) Area km² (MW)	Deep (> 60m) Area km² (MW)	Shallow (0 - 30 m) Area km² (MW)	Transitional (30 - 60m) Area km² (MW)	Deep (> 60m) Area km² (MW)	Shallow (0 - 30 m) Area km² (MW)	Transitional (30 - 60m) Area km² (MW)	Deep (> 60m) Area km² (MW)
7.0 - 7.5	266 (1,331)	236 (1,181)	257 (1,287)	101 (504)	457 (2,284)	4,554 (22,770)	8 (38)	23 (115)	5,537 (27,684)
7.5 - 8.0	239 (1,196)	257 (1,285)	190 (948)	79 (394)	596 (2,978)	3,855 (19,273)	0 (0)	33 (165)	19,616 (98,080)
8.0 - 8.5	125 (626)	178 (891)	282 (1,409)	7 (36)	106 (529)	4,539 (22,695)	0 (0)	0 (0)	17,822 (89,111)
8.5 - 9.0	43 (216)	142 (708)	176 (882)	1 (3)	38 (190)	4,560 (22,799)	0 (0)	0 (0)	17,892 (89,460)
9.0 - 9.5	2 (10)	19 (94)	15 (74)	0 (0)	1 (4)	988 (4,940)	0 (0)	0 (0)	12,160 (60,801)
9.5 - 10.0	0 (0)	6 (30)	14 (69)	0 (0)	0 (0)	656 (3,280)	0 (0)	0 (0)	14,555 (72,774)
>10.0	0 (0)	0 (0)	0 (1)	0 (0)	0 (0)	288 (1,441)	0 (0)	0 (0)	6,638 (33,188)
Total >7.0	676 (3,379)	838 (4,189)	934 (4,670)	187 (937)	1,197 (5,985)	19,440 (97,198)	8 (38)	56 (279)	94,220 (471,098)

45

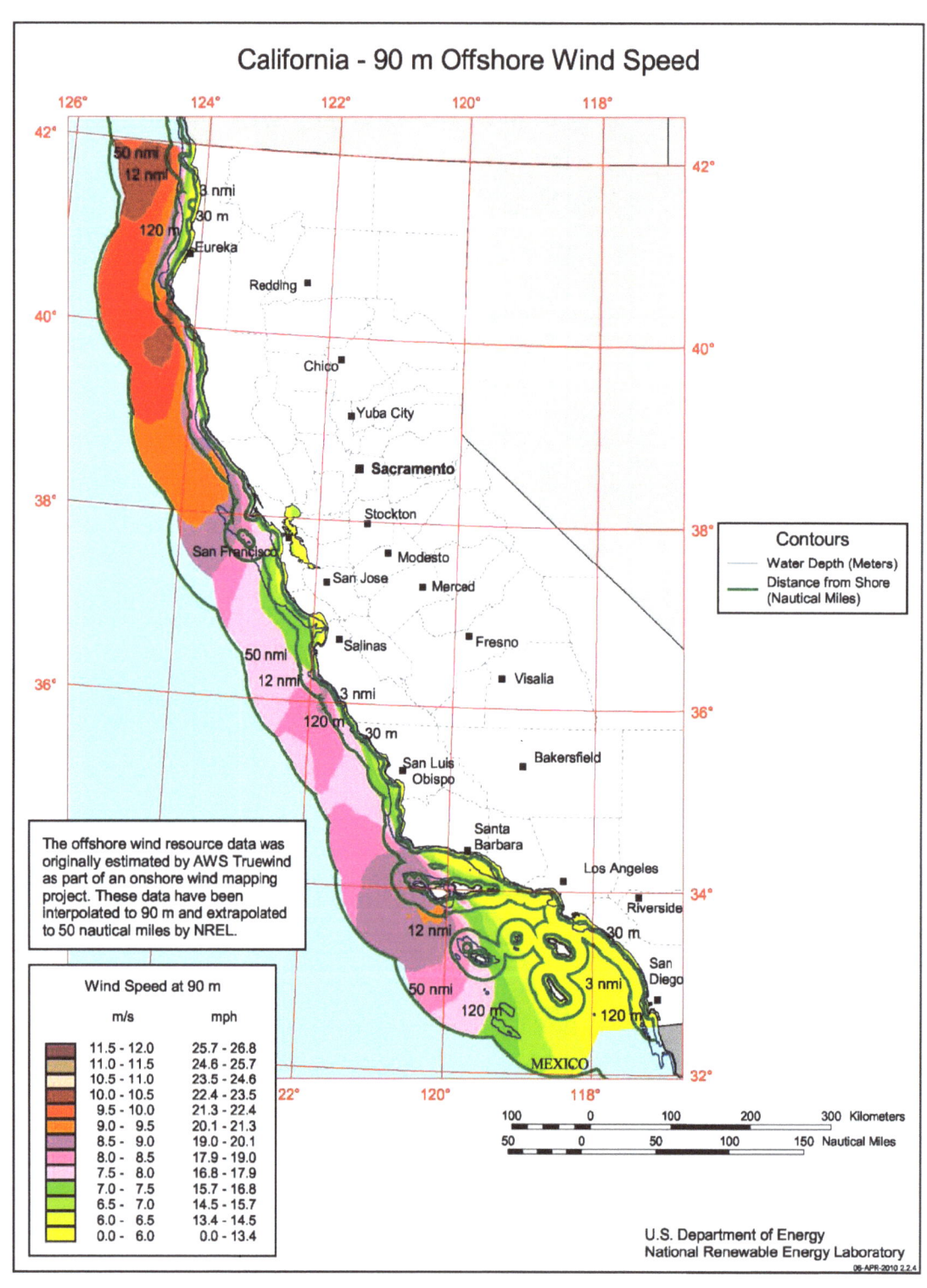

Figure B2. California detailed map

Table B3. Connecticut offshore wind resource by wind speed interval, water depth and distance from shore within 50 nm of shore.

	Distance from Shore (nm)								
	0 – 3			3 – 12			12 – 50		
Depth Category	Shallow (0 – 30 m)	Transitional (30 – 60m)	Deep (> 60m)	Shallow (0 – 30 m)	Transitional (30 – 60m)	Deep (> 60m)	Shallow (0 – 30 m)	Transitional (30 – 60m)	Deep (> 60m)
90 m Wind Speed Interval (m/s)	Area km² (MW)	Area km² (MW)	Area km² (MW)	Area km² (MW)	Area km² (MW)	Area km² (MW)	Area km² (MW)	Area km² (MW)	Area km² (MW)
7.0 – 7.5	500 (2,501)	30 (151)	0 (0)	0 (0)	0 (0)	0 (0)	0 (0)	0 (0)	0 (0)
7.5 – 8.0	617 (3,087)	83 (415)	1 (6)	0 (0)	0 (0)	0 (0)	0 (0)	0 (0)	0 (0)
8.0 – 8.5	35 (173)	5 (25)	0 (2)	0 (0)	0 (0)	0 (0)	0 (0)	0 (0)	0 (0)
8.5 – 9.0	0 (0)	0 (0)	0 (0)	0 (0)	0 (0)	0 (0)	0 (0)	0 (0)	0 (0)
9.0 – 9.5	0 (0)	0 (0)	0 (0)	0 (0)	0 (0)	0 (0)	0 (0)	0 (0)	0 (0)
9.5 – 10.0	0 (0)	0 (0)	0 (0)	0 (0)	0 (0)	0 (0)	0 (0)	0 (0)	0 (0)
>10.0	0 (0)	0 (0)	0 (0)	0 (0)	0 (0)	0 (0)	0 (0)	0 (0)	0 (0)
Total >7.0	1,152 (5,760)	118 (592)	2 (8)	0 (0)	0 (0)	0 (0)	0 (0)	0 (0)	0 (0)

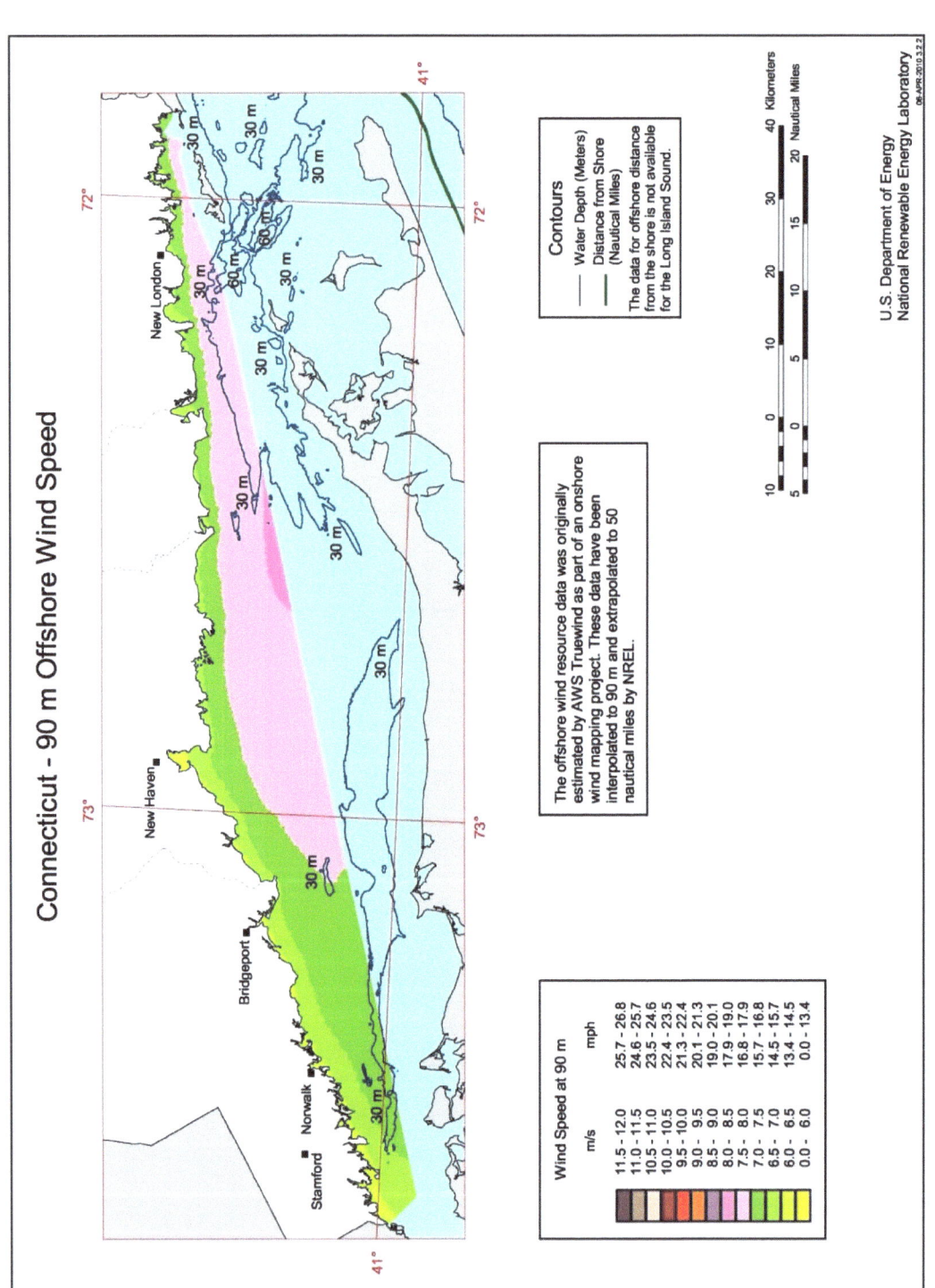

Figure B3. Connecticut detailed map

48

Table B4. Delaware offshore wind resource by wind speed interval, water depth and distance from shore within 50 nm of shore.

	Distance from Shore (nm)								
	0 - 3			3 - 12			12 - 50		
Depth Category	Shallow (0 - 30 m)	Transitional (30 - 60m)	Deep (> 60m)	Shallow (0 - 30 m)	Transitional (30 - 60m)	Deep (> 60m)	Shallow (0 - 30 m)	Transitional (30 - 60m)	Deep (> 60m)
90 m Wind Speed Interval (m/s)	Area km² (MW)	Area km² (MW)	Area km² (MW)	Area km² (MW)	Area km² (MW)	Area km² (MW)	Area km² (MW)	Area km² (MW)	Area km² (MW)
7.0 - 7.5	223 (1,116)	0 (0)	0 (0)	0 (0)	0 (0)	0 (0)	0 (0)	0 (0)	0 (0)
7.5 - 8.0	717 (3,583)	2 (10)	0 (0)	5 (26)	0 (0)	0 (0)	0 (0)	0 (0)	0 (0)
8.0 - 8.5	135 (677)	11 (53)	0 (0)	658 (3,290)	9 (43)	0 (0)	240 (1,202)	9 (44)	0 (0)
8.5 - 9.0	0 (0)	0 (0)	0 (0)	0 (0)	0 (0)	0 (0)	254 (1,270)	677 (3,387)	0 (0)
9.0 - 9.5	0 (0)	0 (0)	0 (0)	0 (0)	0 (0)	0 (0)	0 (0)	0 (0)	0 (0)
9.5 - 10.0	0 (0)	0 (0)	0 (0)	0 (0)	0 (0)	0 (0)	0 (0)	0 (0)	0 (0)
>10.0	0 (0)	0 (0)	0 (0)	0 (0)	0 (0)	0 (0)	0 (0)	0 (0)	0 (0)
Total >7.0	1,075 (5,375)	13 (63)	0 (0)	663 (3,316)	9 (43)	0 (0)	494 (2,472)	686 (3,431)	0 (0)

49

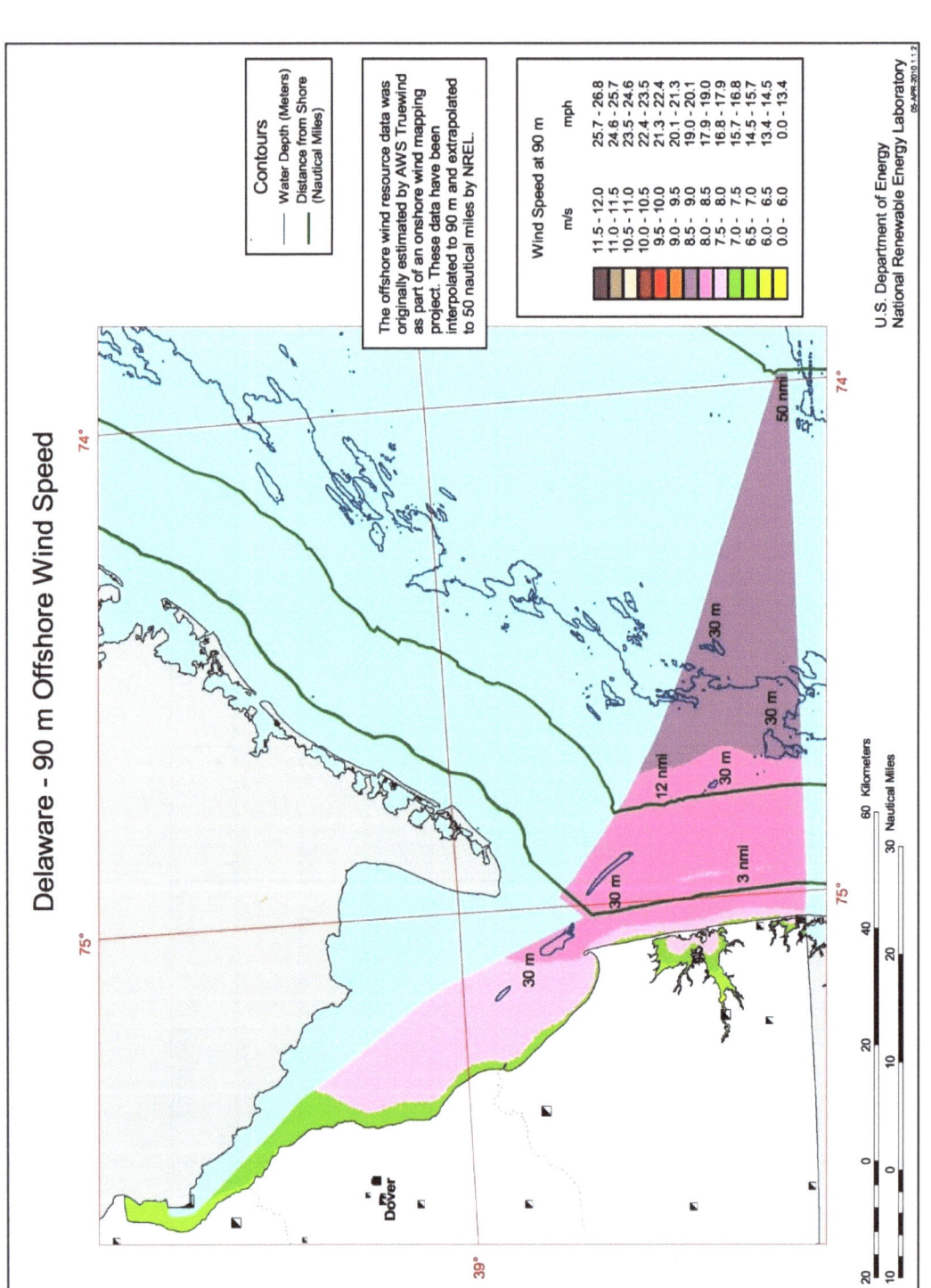

Delaware - 90 m Offshore Wind Speed

Contours
— Water Depth (Meters)
— Distance from Shore (Nautical Miles)

The offshore wind resource data was originally estimated by AWS Truewind as part of an onshore wind mapping project. These data have been interpolated to 90 m and extrapolated to 50 nautical miles by NREL.

Wind Speed at 90 m

m/s	mph
11.5 - 12.0	25.7 - 26.8
11.0 - 11.5	24.6 - 25.7
10.5 - 11.0	23.5 - 24.6
10.0 - 10.5	22.4 - 23.5
9.5 - 10.0	21.3 - 22.4
9.0 - 9.5	20.1 - 21.3
8.5 - 9.0	19.0 - 20.1
8.0 - 8.5	17.9 - 19.0
7.5 - 8.0	16.8 - 17.9
7.0 - 7.5	15.7 - 16.8
6.5 - 7.0	14.5 - 15.7
6.0 - 6.5	13.4 - 14.5
0.0 - 6.0	0.0 - 13.4

U.S. Department of Energy
National Renewable Energy Laboratory

05-APR-2010 1.1.2

Figure B4. Delaware detailed map

Table B5. Georgia offshore wind resource by wind speed interval, water depth and distance from shore within 50 nm of shore.

	Distance from Shore (nm)								
	0 - 3			3 - 12			12 - 50		
Depth Category	Shallow (0 - 30 m)	Transitional (30 - 60m)	Deep (> 60m)	Shallow (0 - 30 m)	Transitional (30 - 60m)	Deep (> 60m)	Shallow (0 - 30 m)	Transitional (30 - 60m)	Deep (> 60m)
90 m Wind Speed Interval (m/s)	Area km² (MW)	Area km² (MW)	Area km² (MW)	Area km² (MW)	Area km² (MW)	Area km² (MW)	Area km² (MW)	Area km² (MW)	Area km² (MW)
7.0 - 7.5	547 (2,737)	0 (0)	0 (0)	2,162 (10,811)	0 (0)	0 (0)	1,111 (5,553)	0 (0)	0 (0)
7.5 - 8.0	85 (426)	0 (0)	0 (0)	530 (2,648)	0 (0)	0 (0)	5,204 (26,021)	1,922 (9,610)	0 (0)
8.0 - 8.5	0 (0)	0 (0)	0 (0)	0 (0)	0 (0)	0 (0)	4 (19)	520 (2,598)	0 (0)
8.5 - 9.0	0 (0)	0 (0)	0 (0)	0 (0)	0 (0)	0 (0)	0 (0)	0 (0)	0 (0)
9.0 - 9.5	0 (0)	0 (0)	0 (0)	0 (0)	0 (0)	0 (0)	0 (0)	0 (0)	0 (0)
9.5 - 10.0	0 (0)	0 (0)	0 (0)	0 (0)	0 (0)	0 (0)	0 (0)	0 (0)	0 (0)
>10.0	0 (0)	0 (0)	0 (0)	0 (0)	0 (0)	0 (0)	0 (0)	0 (0)	0 (0)
Total >7.0	633 (3,164)	0 (0)	0 (0)	2,692 (13,459)	0 (0)	0 (0)	6,319 (31,594)	2,442 (12,208)	0 (0)

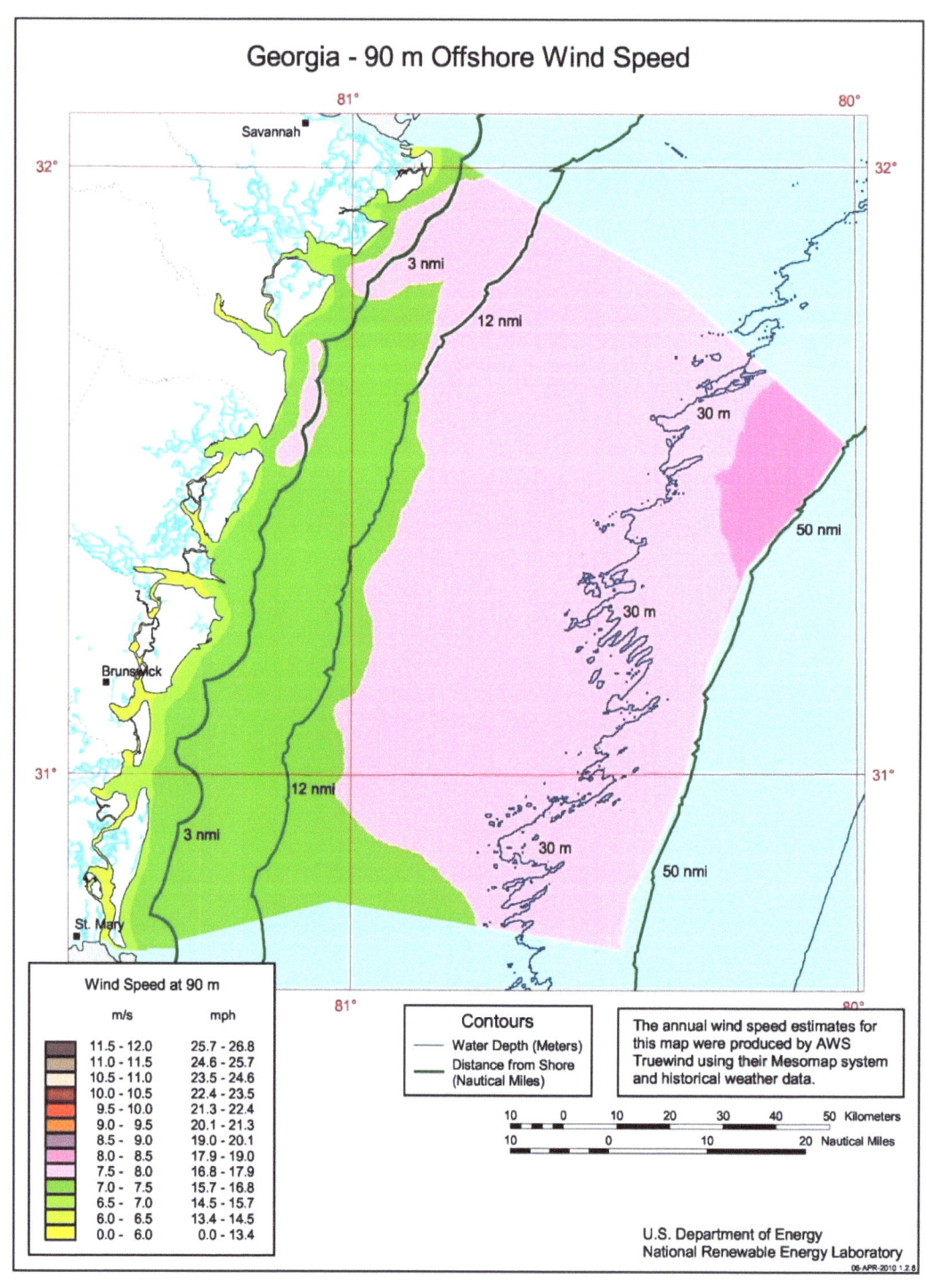

Figure B5. Georgia detailed map

52

Table B6. Hawaii offshore wind resource by wind speed interval, water depth and distance from shore within 50 nm of shore.

Depth Category 90 m Wind Speed Interval (m/s)	Distance from Shore (nm)								
	0 - 3			3 - 12			12 - 50		
	Shallow (0 - 30 m) Area km² (MW)	Transitional (30 - 60m) Area km² (MW)	Deep (> 60m) Area km² (MW)	Shallow (0 - 30 m) Area km² (MW)	Transitional (30 - 60m) Area km² (MW)	Deep (> 60m) Area km² (MW)	Shallow (0 - 30 m) Area km² (MW)	Transitional (30 - 60m) Area km² (MW)	Deep (> 60m) Area km² (MW)
7.0 - 7.5	111 (557)	97 (486)	2,631 (13,157)	0 (0)	43 (213)	2,213 (11,067)	44 (222)	117 (584)	13,615 (68,077)
7.5 - 8.0	66 (328)	107 (535)	2,404 (12,020)	0 (0)	145 (725)	5,052 (25,258)	7 (34)	251 (1,254)	34,268 (171,338)
8.0 - 8.5	92 (461)	115 (574)	2,364 (11,822)	0 (0)	15 (75)	4,756 (23,778)	0 (1)	0 (1)	25,700 (128,498)
8.5 - 9.0	66 (328)	67 (335)	2,105 (10,526)	0 (0)	0 (0)	2,626 (13,128)	0 (0)	0 (0)	9,050 (45,250)
9.0 - 9.5	26 (129)	39 (195)	996 (4,978)	0 (0)	0 (0)	1,853 (9,265)	0 (0)	0 (0)	4,866 (24,328)
9.5 - 10.0	22 (110)	40 (199)	666 (3,330)	0 (0)	0 (0)	1,065 (5,324)	0 (0)	0 (0)	4,927 (24,634)
>10.0	26 (132)	71 (353)	1,344 (6,720)	0 (0)	0 (0)	2,028 (10,138)	0 (0)	0 (0)	1,384 (6,918)
Total >7.0	409 (2,045)	535 (2,677)	12,511 (62,553)	0 (0)	203 (1,013)	19,592 (97,958)	51 (257)	368 (1,839)	93,808 (469,041)

53

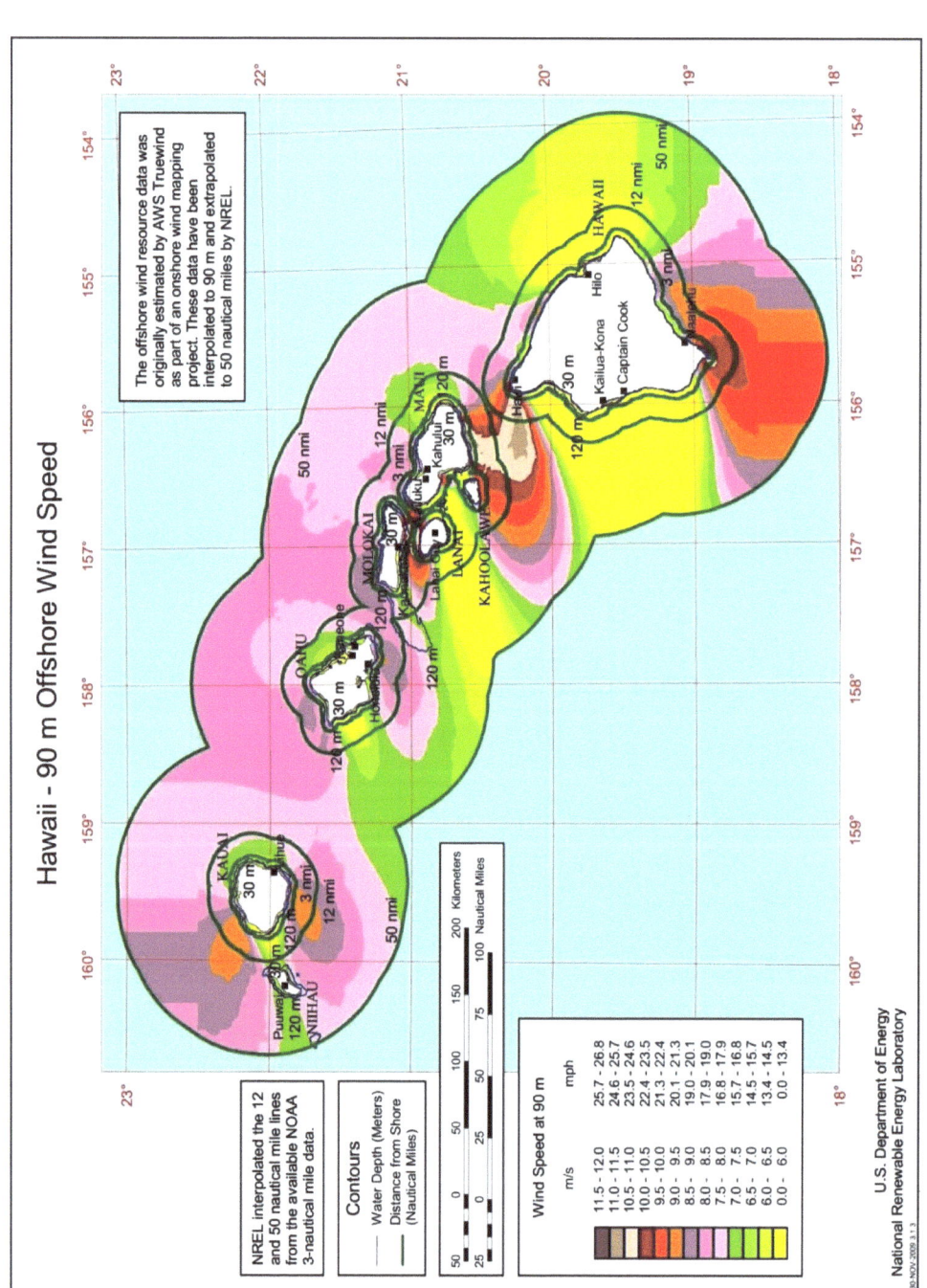

Figure B6. Hawaii detailed map

54

Table B7. Illinois offshore wind resource by wind speed interval, water depth and distance from shore within 50 nm of shore.

Depth Category	Distance from Shore (nm)								
	0 - 3			3 - 12			12 - 50		
90 m Wind Speed Interval (m/s)	Shallow (0 - 30 m) Area km² (MW)	Transitional (30 - 60m) Area km² (MW)	Deep (> 60m) Area km² (MW)	Shallow (0 - 30 m) Area km² (MW)	Transitional (30 - 60m) Area km² (MW)	Deep (> 60m) Area km² (MW)	Shallow (0 - 30 m) Area km² (MW)	Transitional (30 - 60m) Area km² (MW)	Deep (> 60m) Area km² (MW)
7.0 - 7.5	92 (458)	0 (0)	0 (0)	0 (0)	0 (0)	0 (0)	0 (0)	0 (0)	0 (0)
7.5 - 8.0	165 (823)	0 (0)	0 (0)	0 (0)	0 (1)	0 (0)	0 (0)	1 (6)	0 (0)
8.0 - 8.5	244 (1,221)	1 (5)	0 (0)	830 (4,152)	512 (2,561)	141 (704)	7 (35)	426 (2,130)	1,683 (8,414)
8.5 - 9.0	0 (0)	0 (0)	0 (0)	13 (67)	77 (383)	0 (0)	0 (0)	0 (0)	0 (0)
9.0 - 9.5	0 (0)	0 (0)	0 (0)	0 (0)	0 (0)	0 (0)	0 (0)	0 (0)	0 (0)
9.5 - 10.0	0 (0)	0 (0)	0 (0)	0 (0)	0 (0)	0 (0)	0 (0)	0 (0)	0 (0)
>10.0	0 (0)	0 (0)	0 (0)	0 (0)	0 (0)	0 (0)	0 (0)	0 (0)	0 (0)
Total >7.0	500 (2,502)	1 (5)	0 (0)	844 (4,219)	589 (2,944)	141 (704)	7 (35)	427 (2,137)	1,683 (8,414)

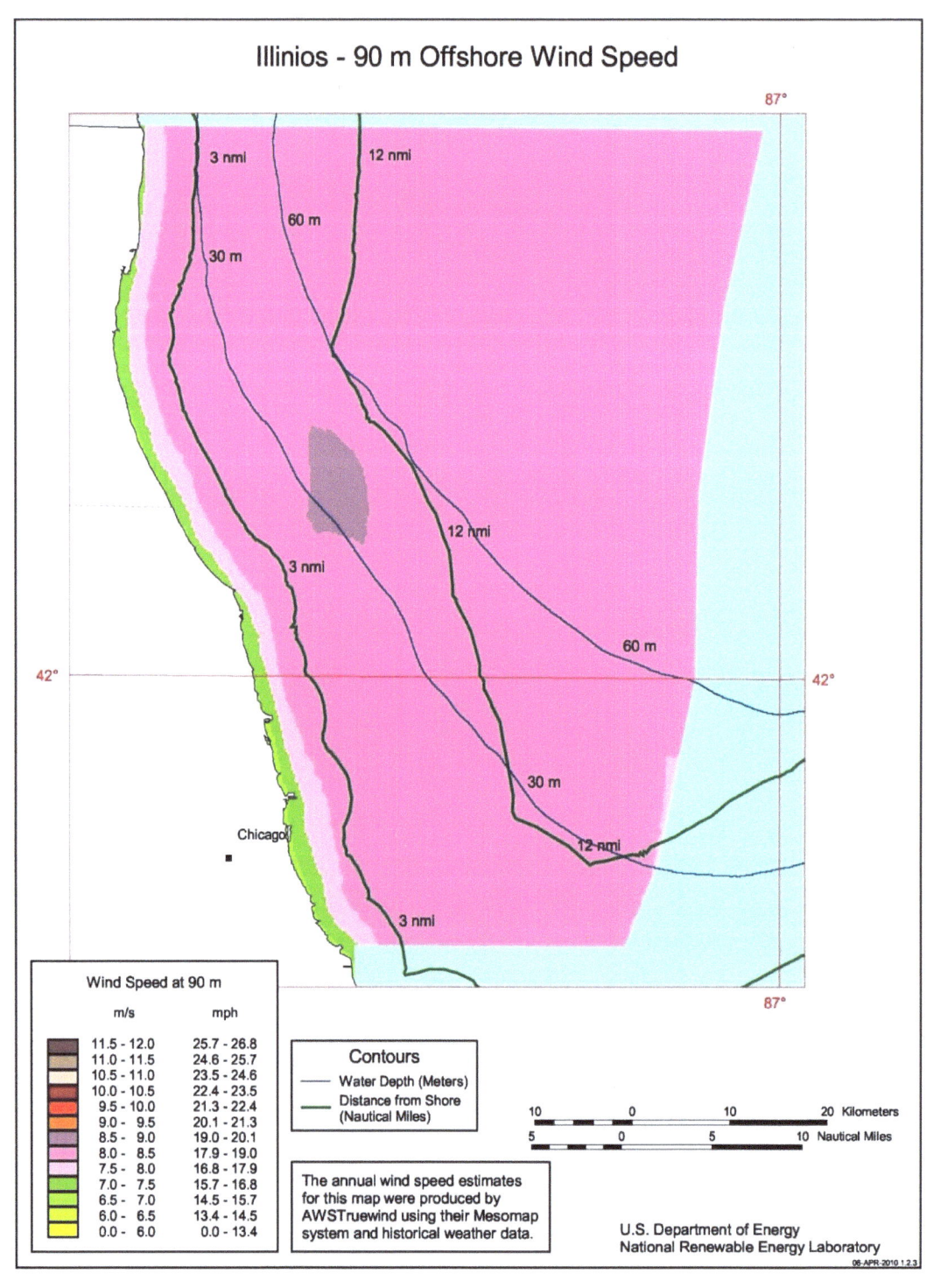

Figure B7. Illinois detailed map

56

Table B8. Indiana offshore wind resource by wind speed interval, water depth and distance from shore within 50 nm of shore.

Depth Category	Distance from Shore (nm)								
	0 - 3			3 - 12			12 - 50		
	Shallow (0 - 30 m)	Transitional (30 - 60m)	Deep (> 60m)	Shallow (0 - 30 m)	Transitional (30 - 60m)	Deep (> 60m)	Shallow (0 - 30 m)	Transitional (30 - 60 m)	Deep (> 60m)
90 m Wind Speed Interval (m/s)	Area km² (MW)	Area km² (MW)	Area km² (MW)	Area km² (MW)	Area km² (MW)	Area km² (MW)	Area km² (MW)	Area km² (MW)	Area km² (MW)
7.0 - 7.5	82 (410)	0 (0)	0 (0)	0 (0)	0 (0)	0 (0)	0 (0)	0 (0)	0 (0)
7.5 - 8.0	154 (769)	0 (0)	0 (0)	63 (313)	0 (0)	0 (0)	0 (0)	0 (0)	0 (0)
8.0 - 8.5	101 (507)	0 (0)	0 (0)	184 (920)	0 (0)	0 (0)	0 (0)	0 (0)	0 (0)
8.5 - 9.0	0 (0)	0 (0)	0 (0)	0 (0)	0 (0)	0 (0)	0 (0)	0 (0)	0 (0)
9.0 - 9.5	0 (0)	0 (0)	0 (0)	0 (0)	0 (0)	0 (0)	0 (0)	0 (0)	0 (0)
9.5 - 10.0	0 (0)	0 (0)	0 (0)	0 (0)	0 (0)	0 (0)	0 (0)	0 (0)	0 (0)
>10.0	0 (0)	0 (0)	0 (0)	0 (0)	0 (0)	0 (0)	0 (0)	0 (0)	0 (0)
Total >7.0	337 (1,687)	0 (0)	0 (0)	247 (1,233)	0 (0)	0 (0)	0 (0)	0 (0)	0 (0)

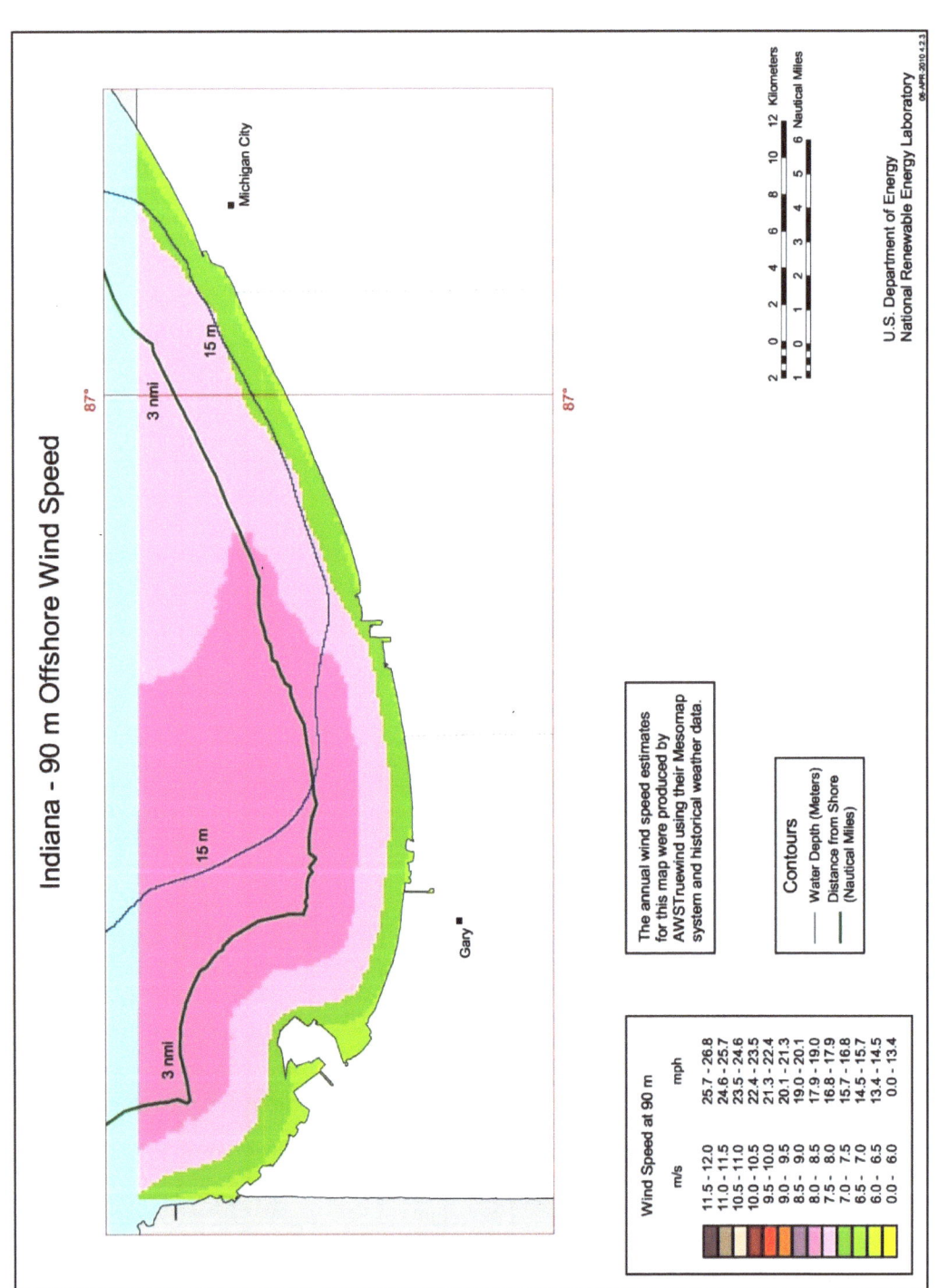

Figure B8. Indiana detailed map

58

Table B9. Louisiana offshore wind resource by wind speed interval, water depth and distance from shore within 50 nm of shore.

	Distance from Shore (nm)								
	0 - 3			3 - 12			12 - 50		
Depth Category	Shallow (0 - 30 m)	Transitional (30 - 60m)	Deep (> 60m)	Shallow (0 - 30 m)	Transitional (30 - 60m)	Deep (> 60m)	Shallow (0 - 30 m)	Transitional (30 - 60m)	Deep (> 60m)
90 m Wind Speed Interval (m/s)	Area km² (MW)	Area km² (MW)	Area km² (MW)	Area km² (MW)	Area km² (MW)	Area km² (MW)	Area km² (MW)	Area km² (MW)	Area km² (MW)
7.0 - 7.5	7,760 (38,798)	29 (143)	0 (0)	7,825 (39,126)	643 (3,217)	1,460 (7,302)	11,164 (55,819)	5,479 (27,396)	13,682 (68,412)
7.5 - 8.0	155 (776)	0 (0)	0 (0)	1,625 (8,123)	0 (0)	0 (0)	8,170 (40,850)	2,229 (11,143)	2,854 (14,271)
8.0 - 8.5	0 (0)	0 (0)	0 (0)	0 (0)	0 (0)	0 (0)	0 (0)	0 (0)	0 (0)
8.5 - 9.0	0 (0)	0 (0)	0 (0)	0 (0)	0 (0)	0 (0)	0 (0)	0 (0)	0 (0)
9.0 - 9.5	0 (0)	0 (0)	0 (0)	0 (0)	0 (0)	0 (0)	0 (0)	0 (0)	0 (0)
9.5 - 10.0	0 (0)	0 (0)	0 (0)	0 (0)	0 (0)	0 (0)	0 (0)	0 (0)	0 (0)
>10.0	0 (0)	0 (0)	0 (0)	0 (0)	0 (0)	0 (0)	0 (0)	0 (0)	0 (0)
Total >7.0	7,915 (39,574)	29 (143)	0 (0)	9,450 (47,249)	643 (3,217)	1,460 (7,302)	19,334 (96,669)	7,708 (38,539)	16,537 (82,683)

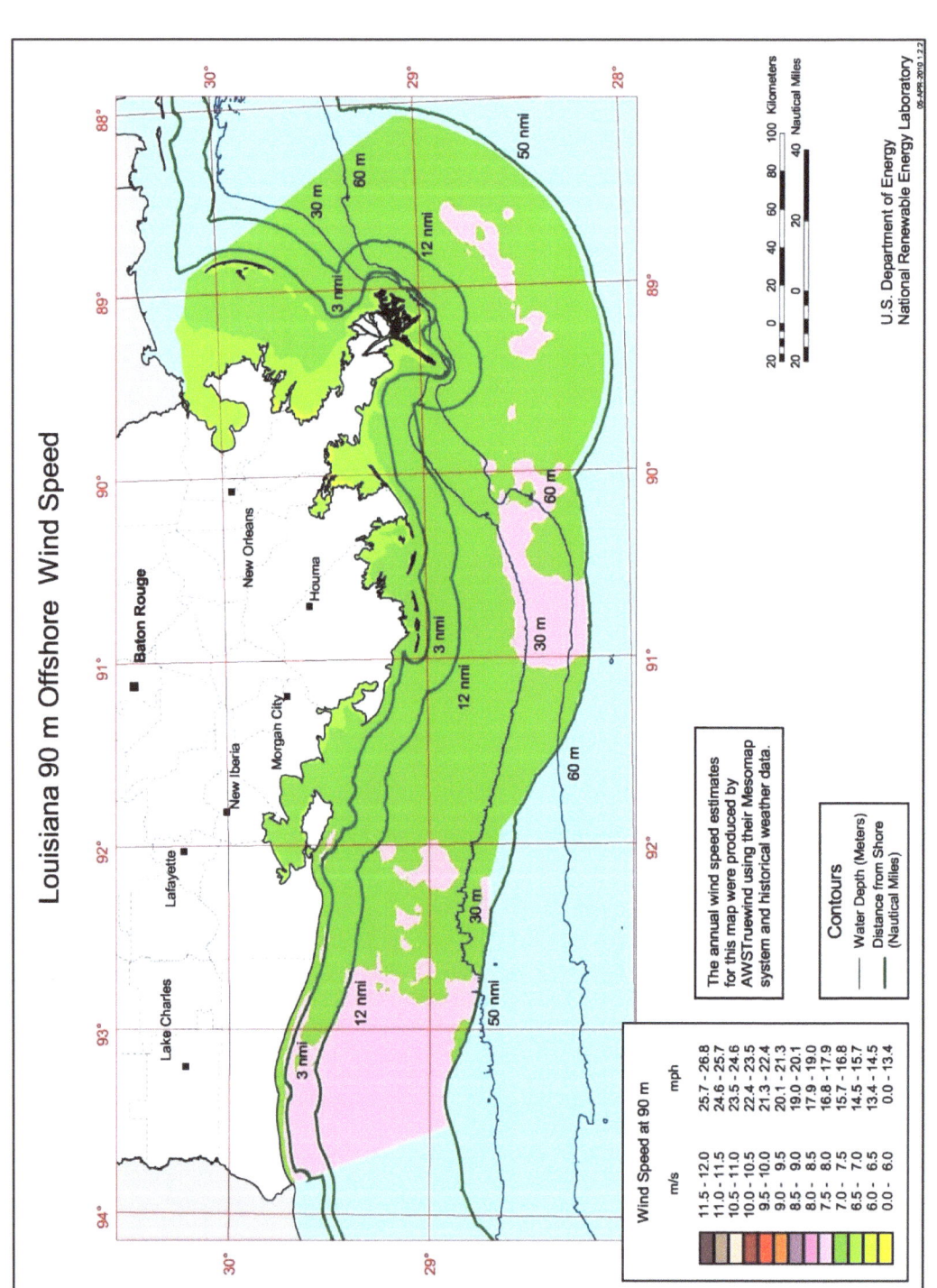

Figure B9. Louisiana detailed map

60

Table B10. Maine offshore wind resource by wind speed interval, water depth and distance from shore within 50 nm of shore.

90 m Wind Speed Interval (m/s)	Distance from Shore (nm)								
	0 - 3			3 - 12			12 - 50		
	Shallow (0 - 30 m) Area km² (MW)	Transitional (30 - 60m) Area km² (MW)	Deep (> 60m) Area km² (MW)	Shallow (0 - 30 m) Area km² (MW)	Transitional (30 - 60m) Area km² (MW)	Deep (> 60m) Area km² (MW)	Shallow (0 - 30 m) Area km² (MW)	Transitional (30 - 60m) Area km² (MW)	Deep (> 60m) Area km² (MW)
7.0 - 7.5	787 (3,935)	91 (456)	12 (59)	8 (39)	5 (24)	4 (18)	0 (0)	0 (0)	0 (0)
7.5 - 8.0	797 (3,986)	285 (1,427)	19 (97)	7 (33)	20 (98)	14 (70)	0 (0)	0 (0)	0 (0)
8.0 - 8.5	777 (3,885)	441 (2,204)	74 (371)	63 (317)	386 (1,928)	235 (1,173)	0 (0)	0 (0)	0 (0)
8.5 - 9.0	513 (2,567)	614 (3,070)	158 (788)	18 (91)	219 (1,095)	1,402 (7,010)	0 (0)	0 (0)	407 (2,034)
9.0 - 9.5	142 (711)	390 (1,950)	309 (1,546)	26 (129)	469 (2,345)	3,504 (17,520)	0 (0)	58 (289)	3,531 (17,655)
9.5 - 10.0	6 (28)	25 (124)	42 (211)	1 (5)	38 (191)	1,460 (7,299)	0 (0)	7 (37)	13,906 (69,528)
>10.0	0 (0)	0 (0)	0 (0)	0 (0)	0 (0)	42 (208)	0 (0)	0 (0)	0 (0)
Total >7.0	3,022 (15,111)	1,846 (9,231)	615 (3,073)	123 (615)	1,136 (5,682)	6,659 (33,297)	0 (0)	65 (326)	17,843 (89,217)

61

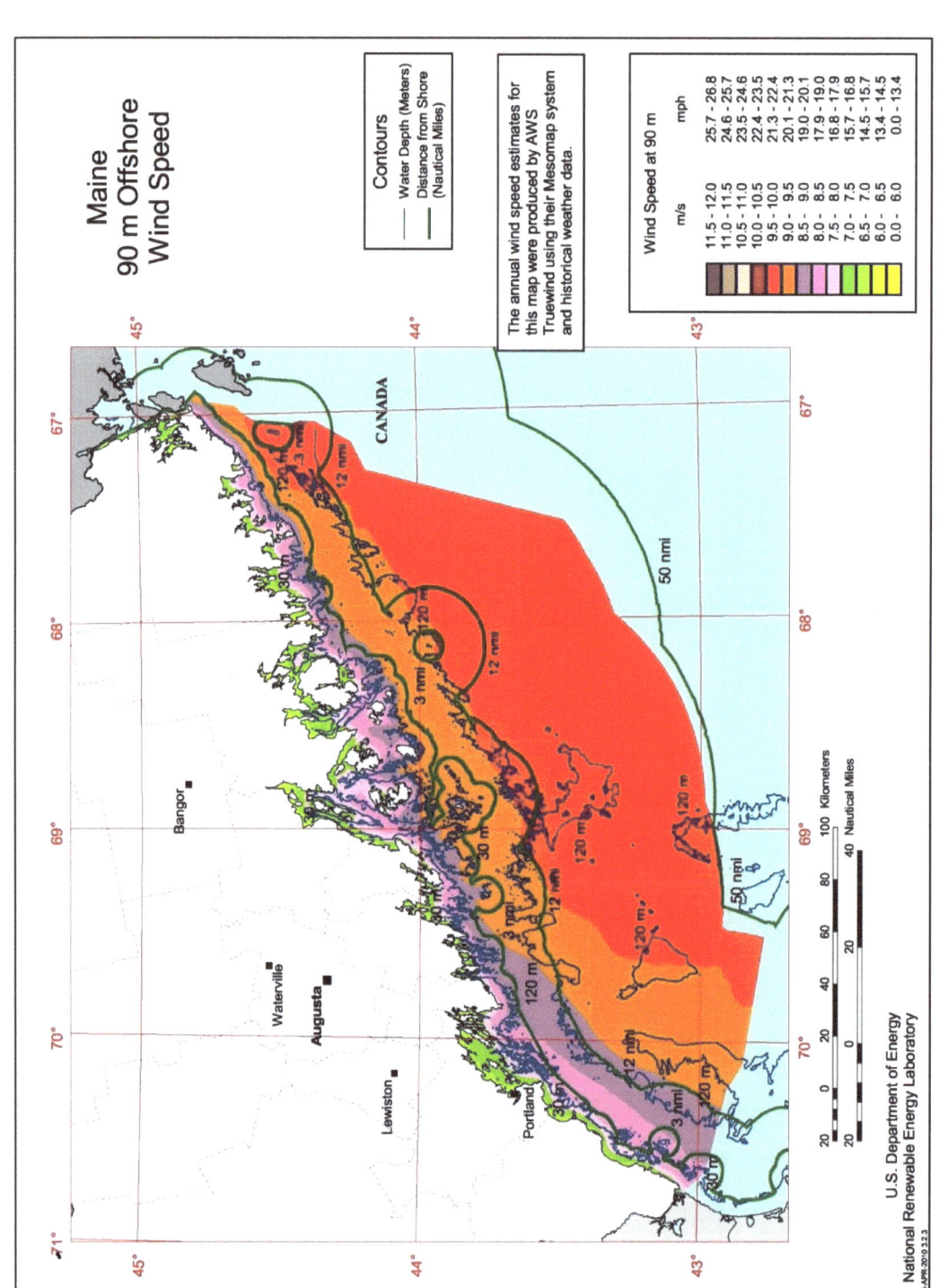

Figure B10. Maine detailed map

62

Table B11. Maryland offshore wind resource by wind speed interval, water depth and distance from shore within 50 nm of shore.

90 m Wind Speed Interval (m/s)	Distance from Shore (nm)								
	0 – 3			3 – 12			12 – 50		
	Shallow (0 – 30 m) Area km² (MW)	Transitional (30 – 60m) Area km² (MW)	Deep (> 60m) Area km² (MW)	Shallow (0 – 30 m) Area km² (MW)	Transitional (30 – 60m) Area km² (MW)	Deep (> 60m) Area km² (MW)	Shallow (0 – 30 m) Area km² (MW)	Transitional (30 – 60m) Area km² (MW)	Deep (> 60m) Area km² (MW)
7.0 - 7.5	2,175 (10,877)	17 (83)	0 (0)	0 (0)	0 (0)	0 (0)	0 (0)	0 (0)	0 (0)
7.5 - 8.0	1,923 (9,613)	14 (70)	0 (0)	10 (49)	0 (0)	0 (0)	0 (0)	0 (0)	0 (0)
8.0 - 8.5	163 (817)	0 (0)	0 (0)	924 (4,620)	0 (0)	0 (0)	435 (2,177)	17 (86)	0 (0)
8.5 - 9.0	0 (0)	0 (0)	0 (0)	0 (0)	0 (0)	0 (0)	531 (2,656)	3,211 (16,053)	1,336 (6,681)
9.0 - 9.5	0 (0)	0 (0)	0 (0)	0 (0)	0 (0)	0 (0)	0 (0)	0 (0)	0 (0)
9.5 - 10.0	0 (0)	0 (0)	0 (0)	0 (0)	0 (0)	0 (0)	0 (0)	0 (0)	0 (0)
>10.0	0 (0)	0 (0)	0 (0)	0 (0)	0 (0)	0 (0)	0 (0)	0 (0)	0 (0)
Total >7.0	4,261 (21,306)	31 (153)	0 (0)	934 (4,670)	0 (0)	0 (0)	967 (4,833)	3,228 (16,139)	1,336 (6,681)

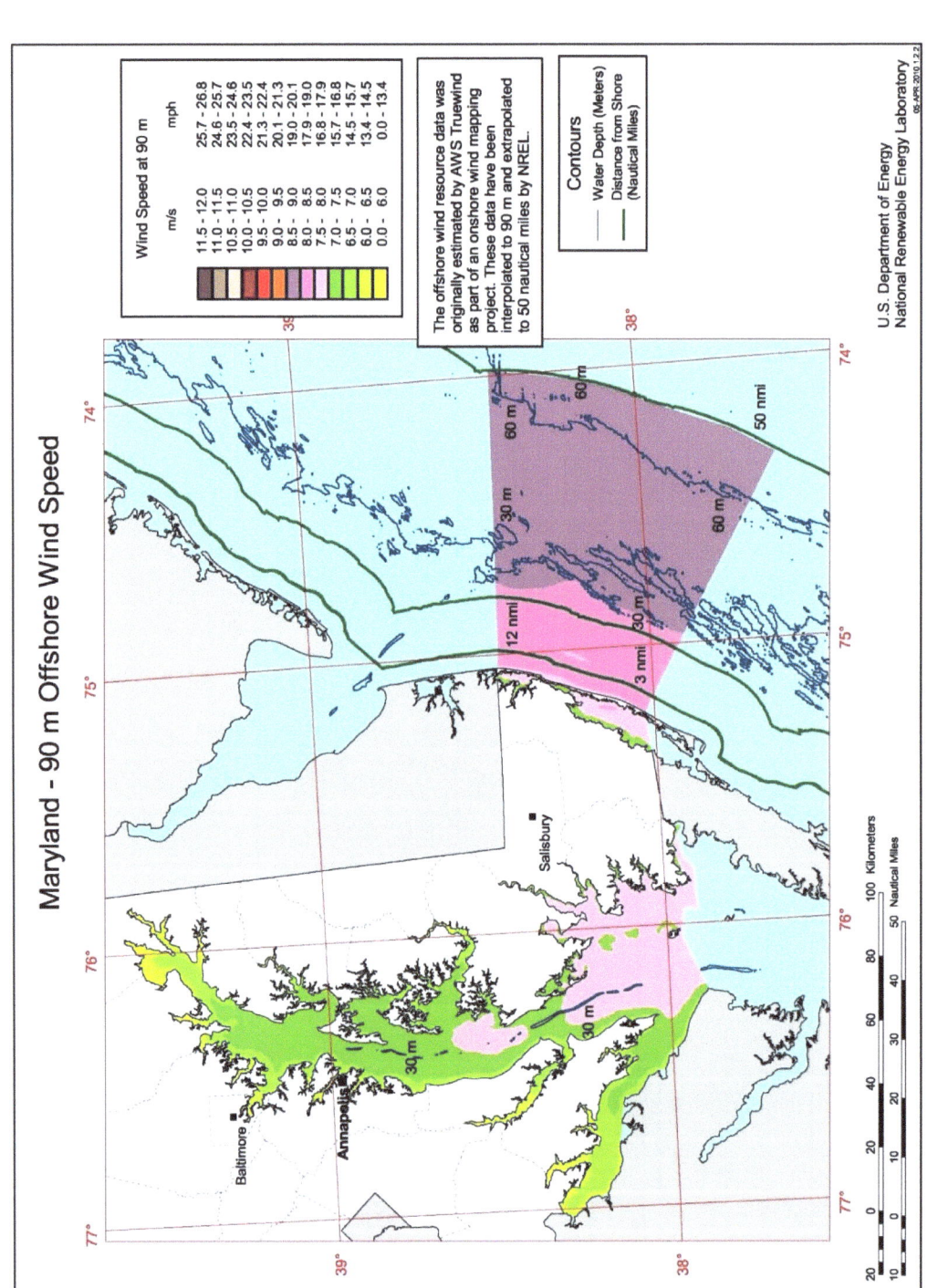

Maryland - 90 m Offshore Wind Speed

Wind Speed at 90 m

m/s	mph
11.5 - 12.0	25.7 - 26.8
11.0 - 11.5	24.6 - 25.7
10.5 - 11.0	23.5 - 24.6
10.0 - 10.5	22.4 - 23.5
9.5 - 10.0	21.3 - 22.4
9.0 - 9.5	20.1 - 21.3
8.5 - 9.0	19.0 - 20.1
8.0 - 8.5	17.9 - 19.0
7.5 - 8.0	16.8 - 17.9
7.0 - 7.5	15.7 - 16.8
6.5 - 7.0	14.5 - 15.7
6.0 - 6.5	13.4 - 14.5
0.0 - 6.0	0.0 - 13.4

The offshore wind resource data was originally estimated by AWS Truewind as part of an onshore wind mapping project. These data have been interpolated to 90 m and extrapolated to 50 nautical miles by NREL.

Contours
— Water Depth (Meters)
— Distance from Shore (Nautical Miles)

U.S. Department of Energy
National Renewable Energy Laboratory

05-APR-2010 1.2.2

Figure B11. Maryland detailed map

64

Table B12. Massachusetts offshore wind resource by wind speed interval, water depth and distance from shore within 50 nm of shore.

	Distance from Shore (nm)								
	0 - 3			3 - 12			12 - 50		
Depth Category	Shallow (0 - 30 m)	Transitional (30 - 60m)	Deep (> 60m)	Shallow (0 - 30 m)	Transitional (30 - 60m)	Deep (> 60m)	Shallow (0 - 30 m)	Transitional (30 - 60m)	Deep (> 60m)
90 m Wind Speed Interval (m/s)	Area km² (MW)	Area km² (MW)	Area km² (MW)	Area km² (MW)	Area km² (MW)	Area km² (MW)	Area km² (MW)	Area km² (MW)	Area km² (MW)
7.0 - 7.5	202 (1,008)	0 (0)	0 (0)	0 (0)	0 (0)	0 (0)	0 (0)	0 (0)	0 (0)
7.5 - 8.0	521 (2,607)	5 (23)	0 (0)	0 (0)	0 (0)	0 (0)	0 (0)	0 (0)	0 (0)
8.0 - 8.5	927 (4,637)	327 (1,636)	29 (143)	78 (391)	152 (760)	126 (628)	0 (0)	0 (0)	0 (0)
8.5 - 9.0	1,508 (7,541)	378 (1,890)	13 (63)	315 (1,575)	355 (1,773)	812 (4,061)	11 (57)	24 (118)	190 (952)
9.0 - 9.5	1,137 (5,685)	323 (1,613)	20 (100)	2,697 (13,484)	1,419 (7,093)	1,007 (5,033)	1,690 (8,449)	5,052 (25,259)	7,007 (35,037)
9.5 - 10.0	2 (10)	0 (0)	0 (0)	9 (43)	119 (596)	0 (0)	472 (2,361)	3,460 (17,298)	9,613 (48,063)
>10.0	0 (0)	0 (0)	0 (0)	0 (0)	0 (0)	0 (0)	0 (0)	0 (0)	0 (0)
Total >7.0	4,298 (21,489)	1,033 (5,163)	61 (307)	3,099 (15,493)	2,044 (10,222)	1,944 (9,721)	2,173 (10,867)	8,535 (42,674)	16,810 (84,051)

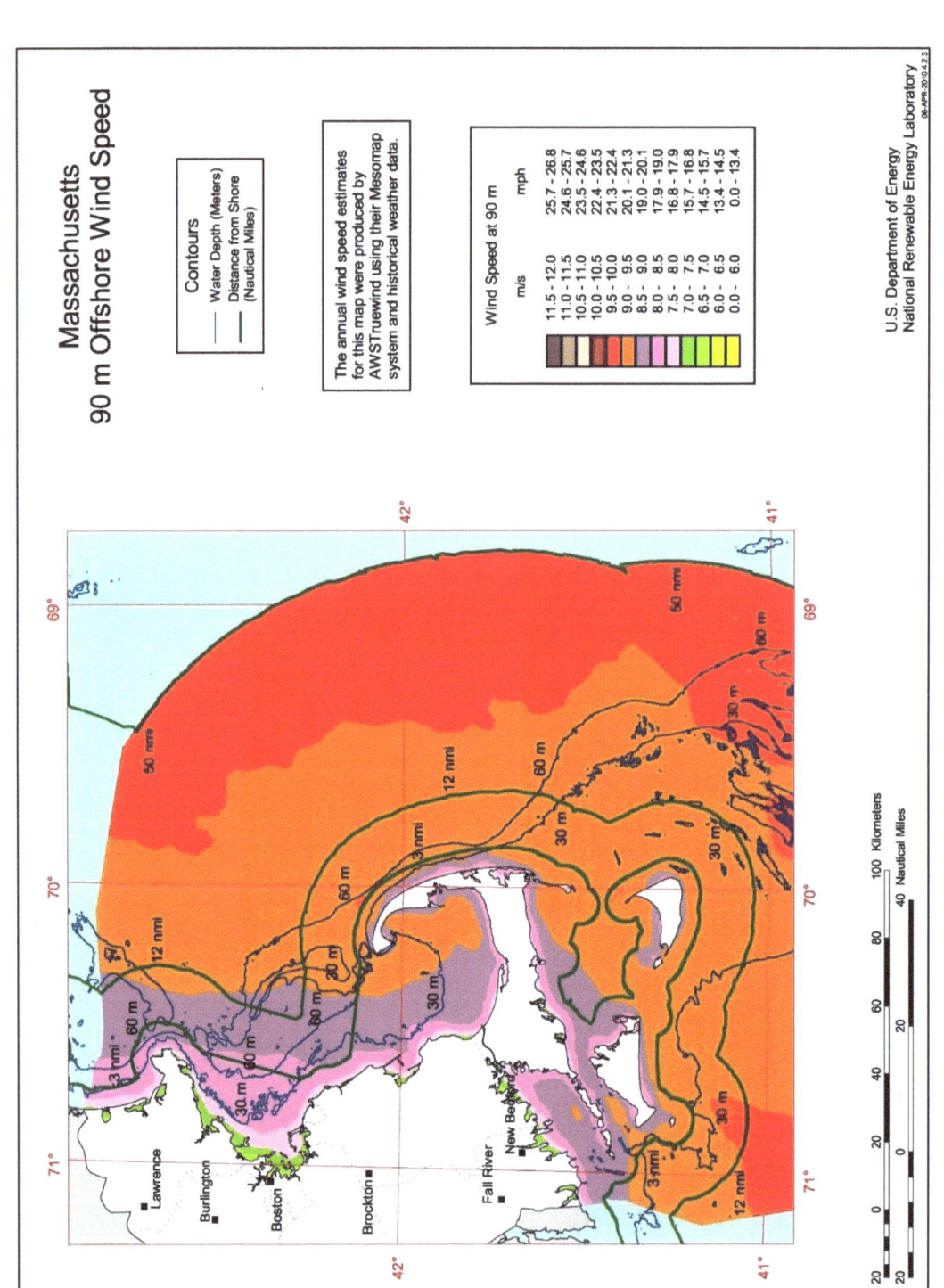

Figure B12. Massachusetts detailed map

66

Table B13. Michigan offshore wind resource by wind speed interval, water depth and distance from shore within 50 nm of shore.

	Distance from Shore (nm)								
	0 - 3			3 - 12			12 - 50		
Depth Category	Shallow (0 - 30 m)	Transitional (30 - 60m)	Deep (> 60m)	Shallow (0 - 30 m)	Transitional (30 - 60m)	Deep (> 60m)	Shallow (0 - 30 m)	Transitional (30 - 60m)	Deep (> 60m)
90 m Wind Speed Interval (m/s)	Area km^2 (MW)	Area km^2 (MW)	Area km^2 (MW)	Area km^2 (MW)	Area km^2 (MW)	Area km^2 (MW)	Area km^2 (MW)	Area km^2 (MW)	Area km^2 (MW)
7.0 - 7.5	3,411 (17,057)	331 (1,653)	244 (1,218)	143 (713)	50 (249)	268 (1,340)	1 (6)	0 (0)	12 (61)
7.5 - 8.0	4,633 (23,163)	1,258 (6,291)	635 (3,177)	3,069 (15,345)	2,916 (14,582)	2,991 (14,955)	14 (71)	108 (542)	2,448 (12,242)
8.0 - 8.5	2,588 (12,940)	1,253 (6,267)	769 (3,845)	2,771 (13,853)	5,890 (29,448)	9,412 (47,059)	31 (155)	876 (4,381)	7,496 (37,482)
8.5 - 9.0	139 (693)	156 (781)	397 (1,983)	145 (726)	551 (2,753)	8,581 (42,904)	118 (592)	1,384 (6,920)	22,834 (114,172)
9.0 - 9.5	0 (0)	0 (0)	0 (0)	0 (0)	0 (0)	70 (350)	115 (573)	250 (1,248)	8,285 (41,423)
9.5 - 10.0	0 (0)	0 (0)	0 (0)	0 (0)	0 (0)	0 (0)	0 (0)	0 (0)	0 (0)
>10.0	0 (0)	0 (0)	0 (0)	0 (0)	0 (0)	0 (0)	0 (0)	0 (0)	0 (0)
Total >7.0	10,771 (53,854)	2,998 (14,991)	2,044 (10,222)	6,127 (30,636)	9,407 (47,033)	21,322 (106,608)	279 (1,397)	2,618 (13,090)	41,076 (205,379)

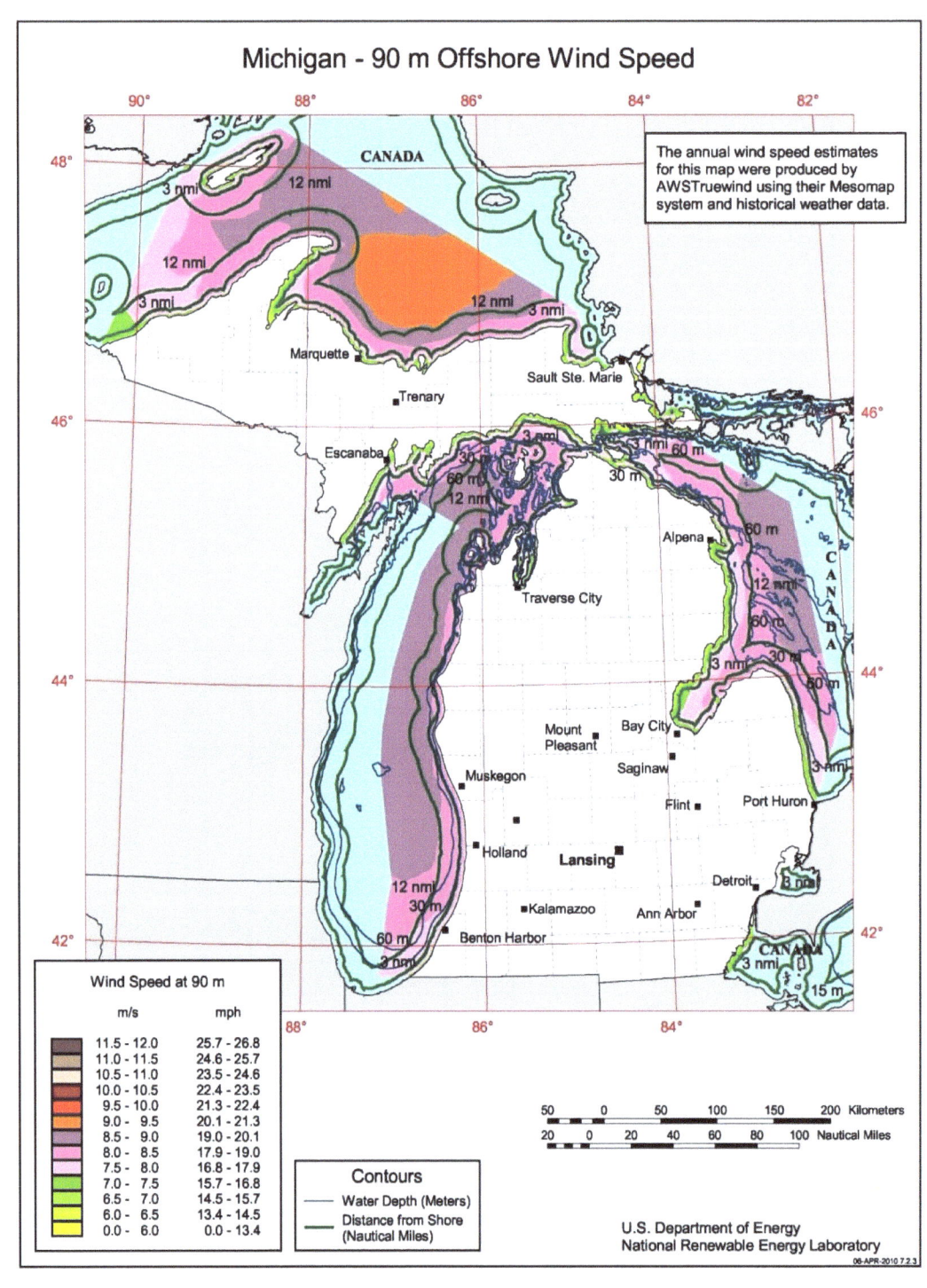

Figure B13. Michigan detailed map

68

Table B14. Minnesota offshore wind resource by wind speed interval, water depth and distance from shore within 50 nm of shore.

	Distance from Shore (nm)								
	0 - 3			3 - 12			12 - 50		
Depth Category	Shallow (0 - 30 m)	Transitional (30 - 60m)	Deep (> 60m)	Shallow (0 - 30 m)	Transitional (30 - 60m)	Deep (> 60m)	Shallow (0 - 30 m)	Transitional (30 - 60m)	Deep (> 60m)
90 m Wind Speed Interval (m/s)	Area km² (MW)	Area km² (MW)	Area km² (MW)	Area km² (MW)	Area km² (MW)	Area km² (MW)	Area km² (MW)	Area km² (MW)	Area km² (MW)
7.0 - 7.5	5 (25)	44 (218)	273 (1,363)	0 (0)	32 (158)	1,718 (8,590)	0 (0)	0 (0)	1,031 (5,155)
7.5 - 8.0	0 (0)	0 (0)	13 (64)	0 (0)	0 (0)	61 (304)	0 (0)	0 (0)	921 (4,603)
8.0 - 8.5	0 (0)	0 (0)	0 (0)	0 (0)	0 (0)	0 (0)	0 (0)	0 (0)	0 (0)
8.5 - 9.0	0 (0)	0 (0)	0 (0)	0 (0)	0 (0)	0 (0)	0 (0)	0 (0)	0 (0)
9.0 - 9.5	0 (0)	0 (0)	0 (0)	0 (0)	0 (0)	0 (0)	0 (0)	0 (0)	0 (0)
9.5 - 10.0	0 (0)	0 (0)	0 (0)	0 (0)	0 (0)	0 (0)	0 (0)	0 (0)	0 (0)
>10.0	0 (0)	0 (0)	0 (0)	0 (0)	0 (0)	0 (0)	0 (0)	0 (0)	0 (0)
Total >7.0	5 (25)	44 (218)	285 (1,427)	0 (0)	32 (158)	1,779 (8,895)	0 (0)	0 (0)	1,952 (9,759)

69

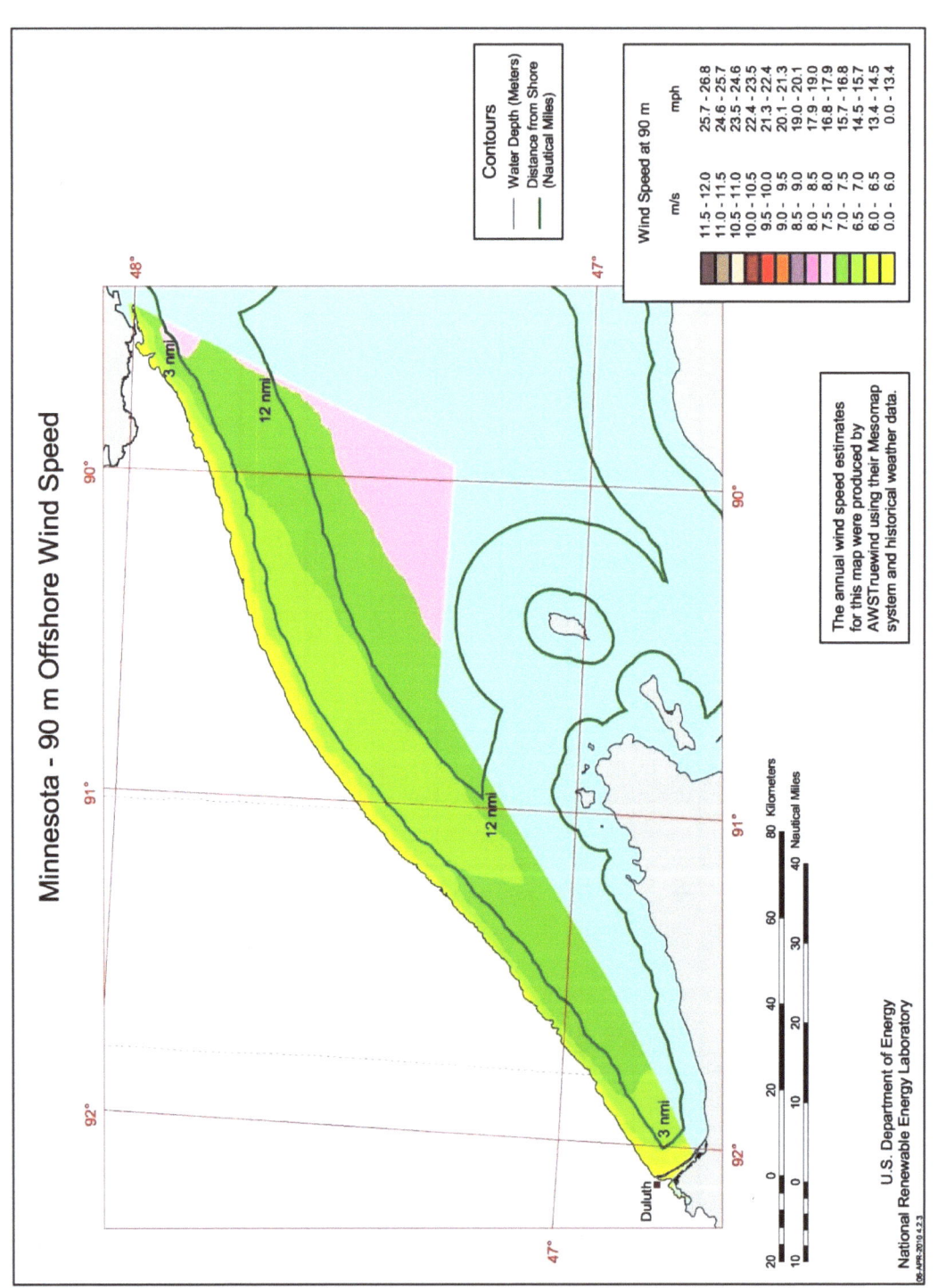

Figure B14. Minnesota detailed tables and maps

70

Table B15. New Hampshire offshore wind resource by wind speed interval, water depth and distance from shore within 50 nm of shore.

Depth Category	Distance from Shore (nm)								
	0 - 3			3 - 12			12 - 50		
	Shallow (0 - 30 m)	Transitional (30 - 60m)	Deep (> 60m)	Shallow (0 - 30 m)	Transitional (30 - 60m)	Deep (> 60m)	Shallow (0 - 30 m)	Transitional (30 - 60m)	Deep (> 60m)
90 m Wind Speed Interval (m/s)	Area km² (MW)	Area km² (MW)	Area km² (MW)	Area km² (MW)	Area km² (MW)	Area km² (MW)	Area km² (MW)	Area km² (MW)	Area km² (MW)
7.0 - 7.5	19 (93)	0 (0)	0 (0)	0 (0)	0 (0)	0 (0)	0 (0)	0 (0)	0 (0)
7.5 - 8.0	46 (229)	0 (0)	0 (0)	0 (0)	0 (0)	0 (0)	0 (0)	0 (0)	0 (0)
8.0 - 8.5	45 (223)	30 (148)	0 (0)	7 (34)	76 (378)	14 (70)	0 (0)	0 (0)	0 (0)
8.5 - 9.0	0 (0)	8 (40)	7 (36)	0 (0)	12 (62)	256 (1,279)	0 (0)	10 (51)	42 (211)
9.0 - 9.5	0 (0)	0 (0)	0 (0)	0 (0)	0 (0)	0 (0)	0 (0)	35 (176)	66 (332)
9.5 - 10.0	0 (0)	0 (0)	0 (0)	0 (0)	0 (0)	0 (0)	0 (0)	0 (0)	0 (0)
>10.0	0 (0)	0 (0)	0 (0)	0 (0)	0 (0)	0 (0)	0 (0)	0 (0)	0 (0)
Total >7.0	109 (545)	38 (188)	7 (36)	7 (35)	88 (439)	270 (1,348)	0 (0)	45 (227)	109 (543)

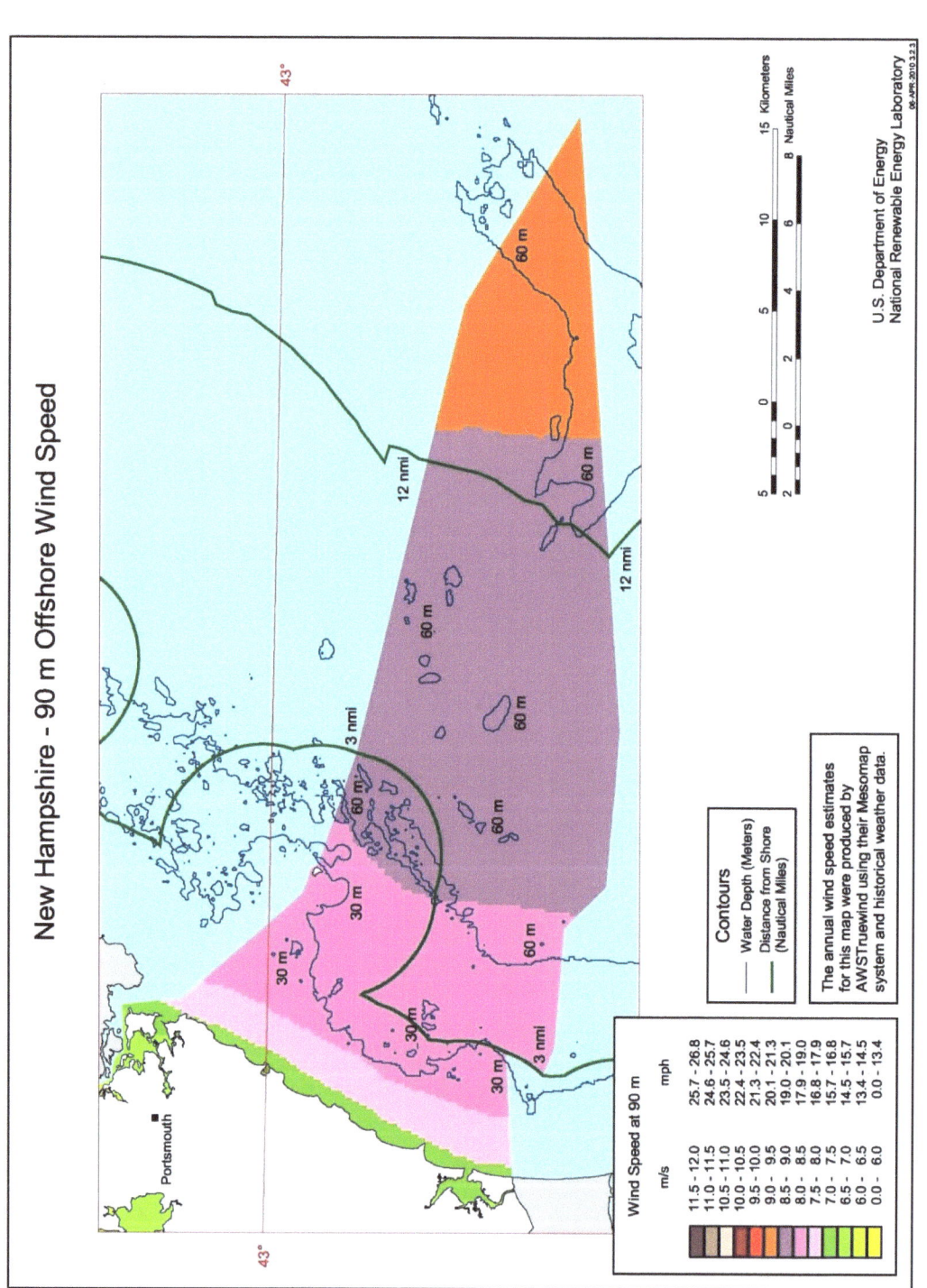

New Hampshire - 90 m Offshore Wind Speed

Wind Speed at 90 m	
m/s	mph
11.5 - 12.0	25.7 - 26.8
11.0 - 11.5	24.6 - 25.7
10.5 - 11.0	23.5 - 24.6
10.0 - 10.5	22.4 - 23.5
9.5 - 10.0	21.3 - 22.4
9.0 - 9.5	20.1 - 21.3
8.5 - 9.0	19.0 - 20.1
8.0 - 8.5	17.9 - 19.0
7.5 - 8.0	16.8 - 17.9
7.0 - 7.5	15.7 - 16.8
6.5 - 7.0	14.5 - 15.7
6.0 - 6.5	13.4 - 14.5
0.0 - 6.0	0.0 - 13.4

Contours
Water Depth (Meters)
Distance from Shore (Nautical Miles)

The annual wind speed estimates for this map were produced by AWSTruewind using their Mesomap system and historical weather data.

U.S. Department of Energy
National Renewable Energy Laboratory
06-APR-2010 3.2.3

Figure B15. New Hampshire detailed map

72

Table B16. New Jersey offshore wind resource by wind speed interval, water depth and distance from shore within 50 nm of shore.

Depth Category 90 m Wind Speed Interval (m/s)	0 - 3 Shallow (0 - 30 m) Area km² (MW)	0 - 3 Transitional (30 - 60m) Area km² (MW)	0 - 3 Deep (> 60m) Area km² (MW)	3 - 12 Shallow (0 - 30 m) Area km² (MW)	3 - 12 Transitional (30 - 60m) Area km² (MW)	3 - 12 Deep (> 60m) Area km² (MW)	12 - 50 Shallow (0 - 30 m) Area km² (MW)	12 - 50 Transitional (30 - 60m) Area km² (MW)	12 - 50 Deep (> 60m) Area km² (MW)
7.0 - 7.5	520 (2,601)	0 (0)	0 (0)	8 (41)	0 (0)	0 (0)	0 (0)	0 (0)	0 (0)
7.5 - 8.0	960 (4,802)	0 (0)	0 (0)	519 (2,593)	28 (140)	1 (4)	0 (0)	0 (0)	0 (0)
8.0 - 8.5	777 (3,885)	0 (0)	0 (0)	2,765 (13,824)	127 (636)	3 (16)	899 (4,497)	372 (1,862)	22 (108)
8.5 - 9.0	13 (66)	0 (0)	0 (0)	215 (1,074)	0 (0)	0 (0)	2,020 (10,098)	10,382 (51,912)	304 (1,519)
9.0 - 9.5	0 (0)	0 (0)	0 (0)	0 (0)	0 (0)	0 (0)	0 (0)	0 (0)	0 (0)
9.5 - 10.0	0 (0)	0 (0)	0 (0)	0 (0)	0 (0)	0 (0)	0 (0)	0 (0)	0 (0)
>10.0	0 (0)	0 (0)	0 (0)	0 (0)	0 (0)	0 (0)	0 (0)	0 (0)	0 (0)
Total >7.0	2,271 (11,353)	0 (0)	0 (0)	3,506 (17,532)	155 (775)	4 (20)	2,919 (14,595)	10,755 (53,774)	325 (1,627)

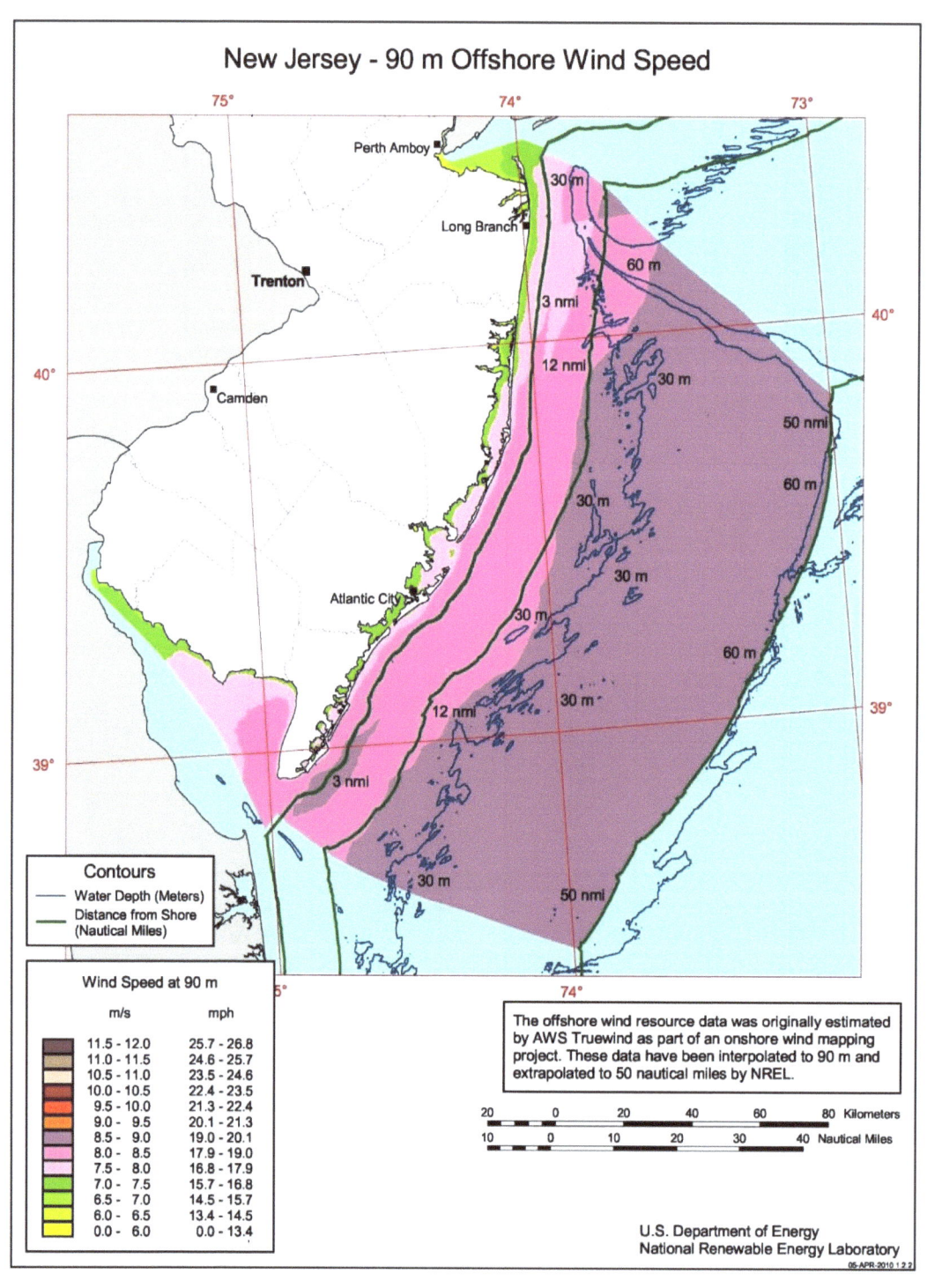

Figure B16. New Jersey detailed map

74

Table B17. New York offshore wind resource by wind speed interval, water depth and distance from shore within 50 nm of shore.

	Distance from Shore (nm)								
	0 - 3			3 - 12			12 - 50		
Depth Category / 90 m Wind Speed Interval (m/s)	Shallow (0 - 30 m) Area km^2 (MW)	Transitional (30 - 60m) Area km^2 (MW)	Deep (> 60m) Area km^2 (MW)	Shallow (0 - 30 m) Area km^2 (MW)	Transitional (30 - 60m) Area km^2 (MW)	Deep (> 60m) Area km^2 (MW)	Shallow (0 - 30 m) Area km^2 (MW)	Transitional (30 - 60m) Area km^2 (MW)	Deep (> 60m) Area km^2 (MW)
7.0 - 7.5	1,048 (5,239)	54 (271)	0 (0)	3 (14)	0 (0)	0 (0)	0 (0)	0 (0)	0 (0)
7.5 - 8.0	1,983 (9,915)	355 (1,774)	12 (58)	473 (2,363)	609 (3,045)	867 (4,334)	0 (0)	0 (0)	60 (300)
8.0 - 8.5	1,329 (6,646)	603 (3,015)	196 (982)	376 (1,878)	536 (2,679)	3,122 (15,610)	42 (211)	13 (63)	2,107 (10,537)
8.5 - 9.0	535 (2,675)	111 (553)	1 (3)	1,043 (5,213)	94 (469)	0 (0)	261 (1,307)	728 (3,639)	104 (522)
9.0 - 9.5	5 (27)	0 (0)	0 (0)	392 (1,959)	1,408 (7,038)	0 (0)	6 (32)	4,926 (24,631)	716 (3,580)
9.5 - 10.0	0 (0)	0 (0)	0 (0)	0 (0)	0 (0)	0 (0)	0 (0)	1,818 (9,088)	3,505 (17,524)
>10.0	0 (0)	0 (0)	0 (0)	0 (0)	0 (0)	0 (0)	0 (0)	0 (0)	0 (0)
Total >7.0	4,901 (24,503)	1,122 (5,612)	209 (1,043)	2,286 (11,428)	2,646 (13,230)	3,989 (19,944)	310 (1,550)	7,484 (37,420)	6,493 (32,463)

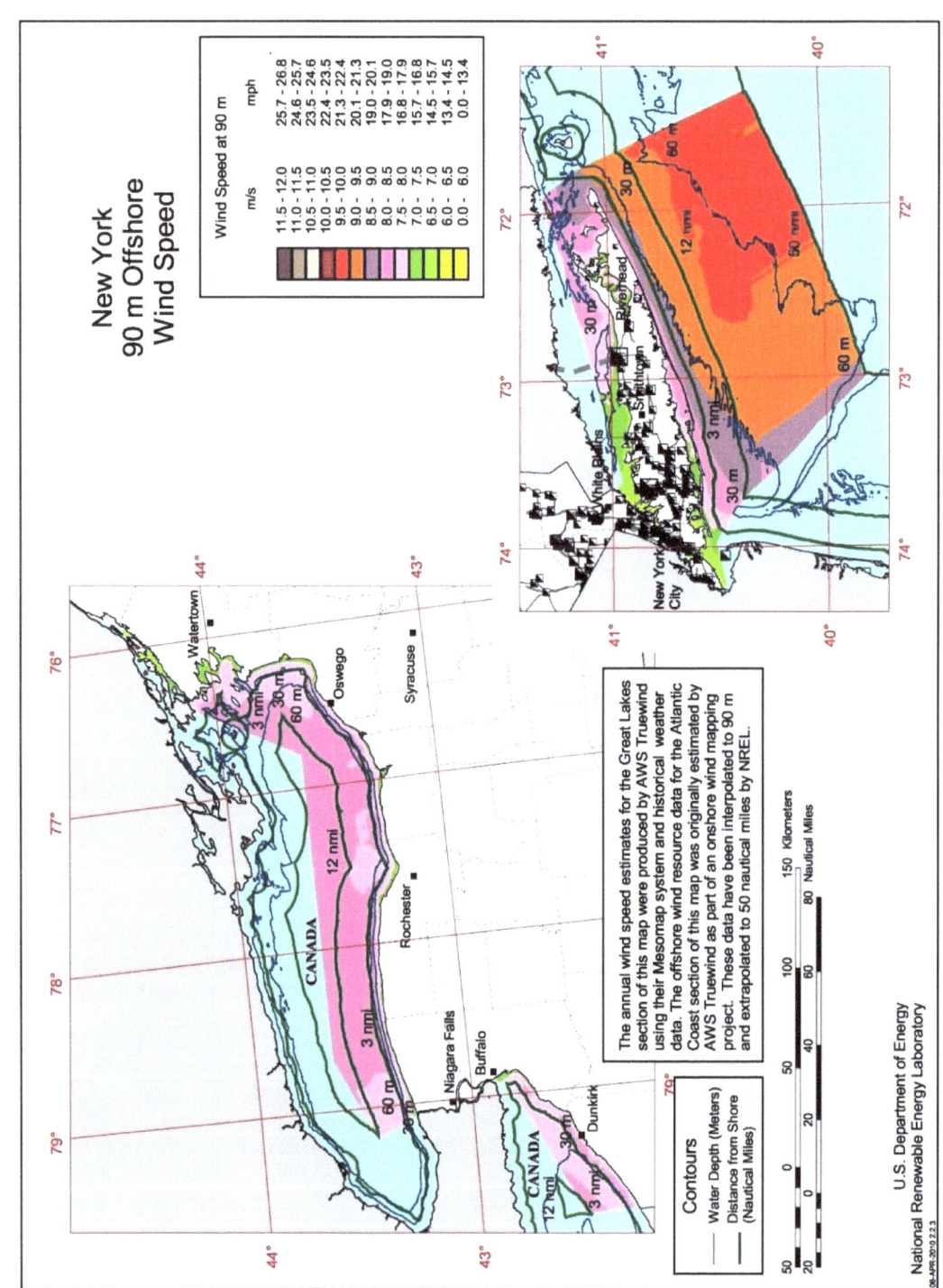

Figure B17. New York detailed map

76

Table B18. North Carolina offshore wind resource by wind speed interval, water depth and distance from shore within 50 nm of shore.

	Distance from Shore (nm)								
	0 - 3			3 - 12			12 - 50		
Depth Category	Shallow (0 - 30 m)	Transitional (30 - 60m)	Deep (> 60m)	Shallow (0 - 30 m)	Transitional (30 - 60m)	Deep (> 60m)	Shallow (0 - 30 m)	Transitional (30 - 60m)	Deep (> 60m)
90 m Wind Speed Interval (m/s)	Area km^2 (MW)	Area km^2 (MW)	Area km^2 (MW)	Area km^2 (MW)	Area km^2 (MW)	Area km^2 (MW)	Area km^2 (MW)	Area km^2 (MW)	Area km^2 (MW)
7.0 - 7.5	1,847 (9,237)	0 (0)	0 (0)	0 (0)	0 (0)	0 (0)	0 (0)	0 (0)	0 (0)
7.5 - 8.0	3,002 (15,010)	0 (0)	0 (0)	1,096 (5,481)	0 (0)	0 (0)	0 (0)	0 (0)	0 (0)
8.0 - 8.5	3,920 (19,600)	0 (1)	0 (0)	4,654 (23,269)	0 (0)	0 (0)	4,035 (20,174)	747 (3,733)	299 (1,496)
8.5 - 9.0	105 (525)	0 (0)	0 (0)	2,964 (14,821)	261 (1,307)	9 (45)	6,382 (31,912)	14,520 (72,598)	15,633 (78,167)
9.0 - 9.5	0 (0)	0 (0)	0 (0)	0 (0)	0 (0)	0 (0)	0 (0)	16 (80)	0 (0)
9.5 - 10.0	0 (0)	0 (0)	0 (0)	0 (0)	0 (0)	0 (0)	0 (0)	0 (0)	0 (0)
>10.0	0 (0)	0 (0)	0 (0)	0 (0)	0 (0)	0 (0)	0 (0)	0 (0)	0 (0)
Total >7.0	8,874 (44,372)	0 (1)	0 (0)	8,714 (43,570)	261 (1,307)	9 (45)	10,417 (52,087)	15,282 (76,411)	15,933 (79,663)

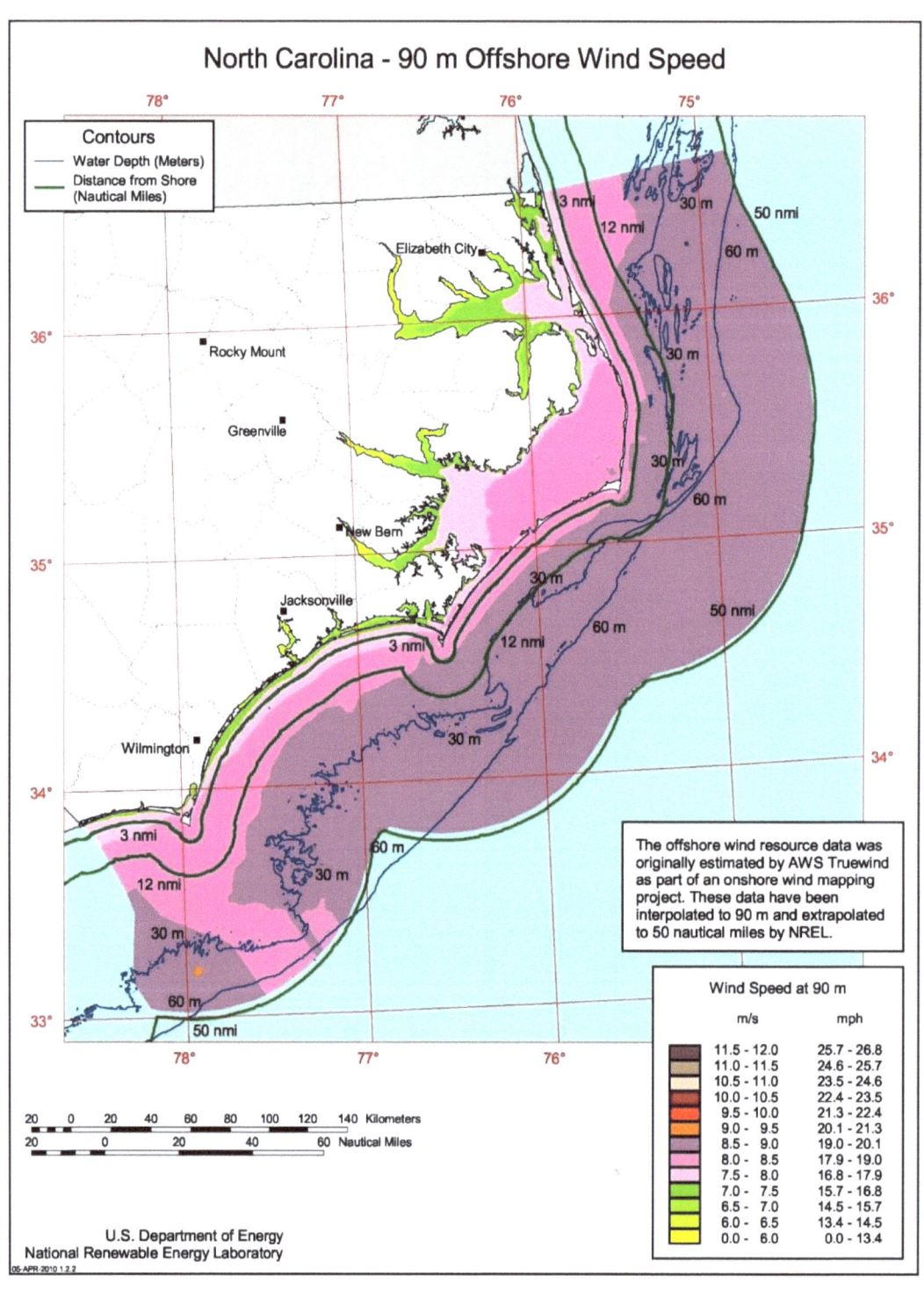

Figure B18. North Carolina detailed map

Table B19. Ohio offshore wind resource by wind speed interval, water depth and distance from shore within 50 nm of shore.

Depth Category	Distance from Shore (nm)								
	0 - 3			3 - 12			12 - 50		
90 m Wind Speed Interval (m/s)	Shallow (0 - 30 m) Area km² (MW)	Transitional (30 - 60m) Area km² (MW)	Deep (> 60m) Area km² (MW)	Shallow (0 - 30 m) Area km² (MW)	Transitional (30 - 60m) Area km² (MW)	Deep (> 60m) Area km² (MW)	Shallow (0 - 30 m) Area km² (MW)	Transitional (30 - 60m) Area km² (MW)	Deep (> 60m) Area km² (MW)
7.0 - 7.5	341 (1,705)	0 (0)	0 (0)	0 (0)	0 (0)	0 (0)	0 (0)	0 (0)	0 (0)
7.5 - 8.0	1,107 (5,534)	0 (0)	0 (0)	1,960 (9,799)	0 (0)	0 (0)	0 (0)	0 (0)	0 (0)
8.0 - 8.5	565 (2,826)	0 (0)	0 (0)	2,193 (10,965)	0 (0)	0 (0)	3,071 (15,356)	0 (0)	0 (0)
8.5 - 9.0	0 (0)	0 (0)	0 (0)	0 (0)	0 (0)	0 (0)	0 (0)	0 (0)	0 (0)
9.0 - 9.5	0 (0)	0 (0)	0 (0)	0 (0)	0 (0)	0 (0)	0 (0)	0 (0)	0 (0)
9.5 - 10.0	0 (0)	0 (0)	0 (0)	0 (0)	0 (0)	0 (0)	0 (0)	0 (0)	0 (0)
>10.0	0 (0)	0 (0)	0 (0)	0 (0)	0 (0)	0 (0)	0 (0)	0 (0)	0 (0)
Total >7.0	2,013 (10,065)	0 (0)	0 (0)	4,153 (20,765)	0 (0)	0 (0)	3,071 (15,356)	0 (0)	0 (0)

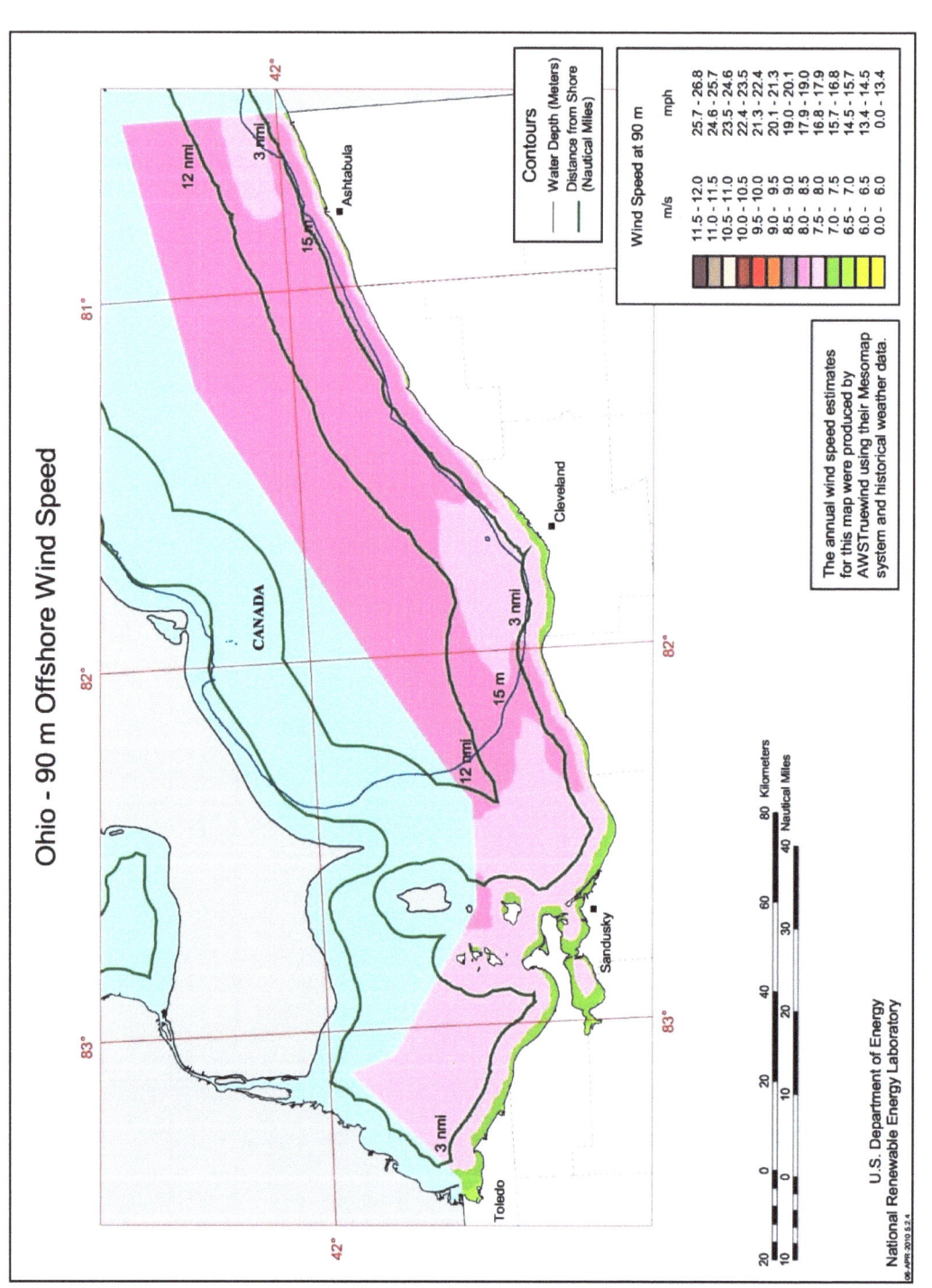

Figure B19. Ohio detailed map

80

Table B20. Oregon offshore wind resource by wind speed interval, water depth and distance from shore within 50 nm of shore.

Depth Category	Distance from Shore (nm)								
	0 - 3			3 - 12			12 - 50		
90 m Wind Speed Interval (m/s)	Shallow (0 - 30 m) Area km² (MW)	Transitional (30 - 60 m) Area km² (MW)	Deep (> 60m) Area km² (MW)	Shallow (0 - 30 m) Area km² (MW)	Transitional (30 - 60 m) Area km² (MW)	Deep (> 60m) Area km² (MW)	Shallow (0 - 30 m) Area km² (MW)	Transitional (30 - 60 m) Area km² (MW)	Deep (> 60m) Area km² (MW)
7.0 - 7.5	356 (1,779)	21 (103)	0 (0)	1 (4)	9 (46)	1 (6)	0 (0)	0 (0)	0 (0)
7.5 - 8.0	523 (2,615)	319 (1,596)	38 (188)	46 (232)	232 (1,159)	335 (1,675)	0 (0)	0 (0)	0 (0)
8.0 - 8.5	198 (991)	277 (1,385)	7 (33)	19 (95)	596 (2,978)	2,558 (12,792)	0 (0)	0 (0)	4,989 (24,947)
8.5 - 9.0	64 (320)	99 (494)	1 (3)	0 (0)	108 (540)	1,967 (9,836)	0 (0)	46 (228)	11,640 (58,201)
9.0 - 9.5	64 (321)	55 (277)	39 (193)	0 (0)	33 (163)	615 (3,074)	0 (0)	0 (0)	6,588 (32,941)
9.5 - 10.0	47 (237)	80 (402)	15 (73)	0 (0)	34 (169)	635 (3,173)	0 (0)	0 (0)	5,255 (26,273)
>10.0	0 (1)	19 (97)	33 (166)	0 (0)	18 (91)	1,369 (6,843)	0 (0)	0 (0)	4,546 (22,730)
Total >7.0	1,253 (6,264)	871 (4,354)	131 (656)	66 (332)	1,029 (5,146)	7,480 (37,399)	0 (0)	46 (228)	33,019 (165,093)

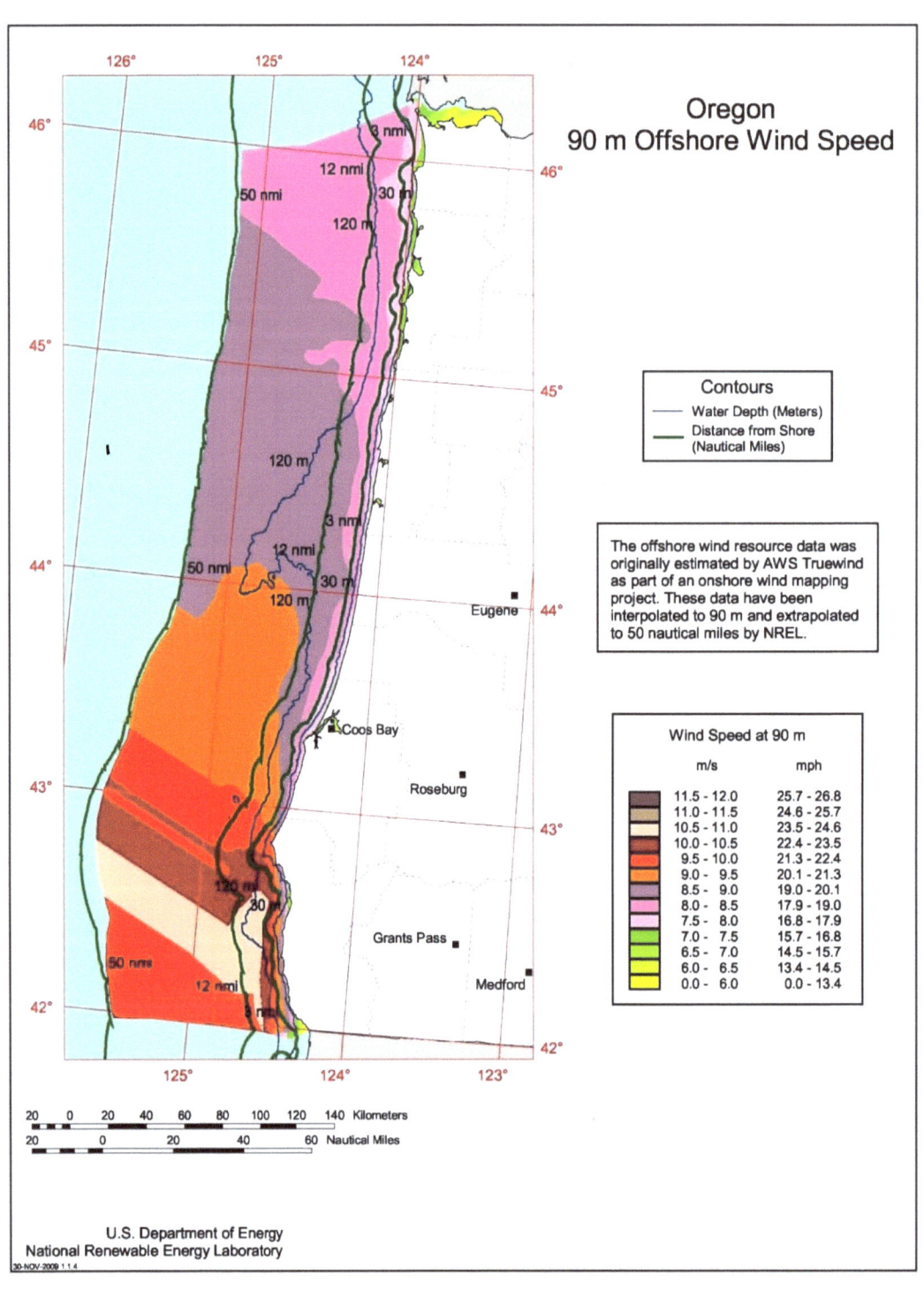

Figure B20. Oregon detailed map

Table B21. Pennsylvania offshore wind resource by wind speed interval, water depth and distance from shore within 50 nm of shore.

	Distance from Shore (nm)								
	0 - 3			3 - 12			12 - 50		
Depth Category	Shallow (0 - 30 m)	Transitional (30 - 60m)	Deep (> 60m)	Shallow (0 - 30 m)	Transitional (30 - 60m)	Deep (> 60m)	Shallow (0 - 30 m)	Transitional (30 - 60m)	Deep (> 60m)
90 m Wind Speed Interval (m/s)	Area km² (MW)	Area km² (MW)	Area km² (MW)	Area km² (MW)	Area km² (MW)	Area km² (MW)	Area km² (MW)	Area km² (MW)	Area km² (MW)
7.0 - 7.5	34 (171)	0 (0)	0 (0)	0 (0)	0 (0)	0 (0)	0 (0)	0 (0)	0 (0)
7.5 - 8.0	113 (567)	0 (0)	0 (0)	53 (266)	44 (221)	0 (0)	0 (0)	0 (0)	0 (0)
8.0 - 8.5	276 (1,381)	1 (4)	0 (0)	751 (3,754)	358 (1,790)	0 (0)	223 (1,114)	70 (351)	0 (0)
8.5 - 9.0	0 (0)	0 (0)	0 (0)	0 (0)	0 (0)	0 (0)	0 (0)	0 (0)	0 (0)
9.0 - 9.5	0 (0)	0 (0)	0 (0)	0 (0)	0 (0)	0 (0)	0 (0)	0 (0)	0 (0)
9.5 - 10.0	0 (0)	0 (0)	0 (0)	0 (0)	0 (0)	0 (0)	0 (0)	0 (0)	0 (0)
>10.0	0 (0)	0 (0)	0 (0)	0 (0)	0 (0)	0 (0)	0 (0)	0 (0)	0 (0)
Total >7.0	424 (2,119)	1 (4)	0 (0)	804 (4,020)	402 (2,012)	0 (0)	223 (1,114)	70 (351)	0 (0)

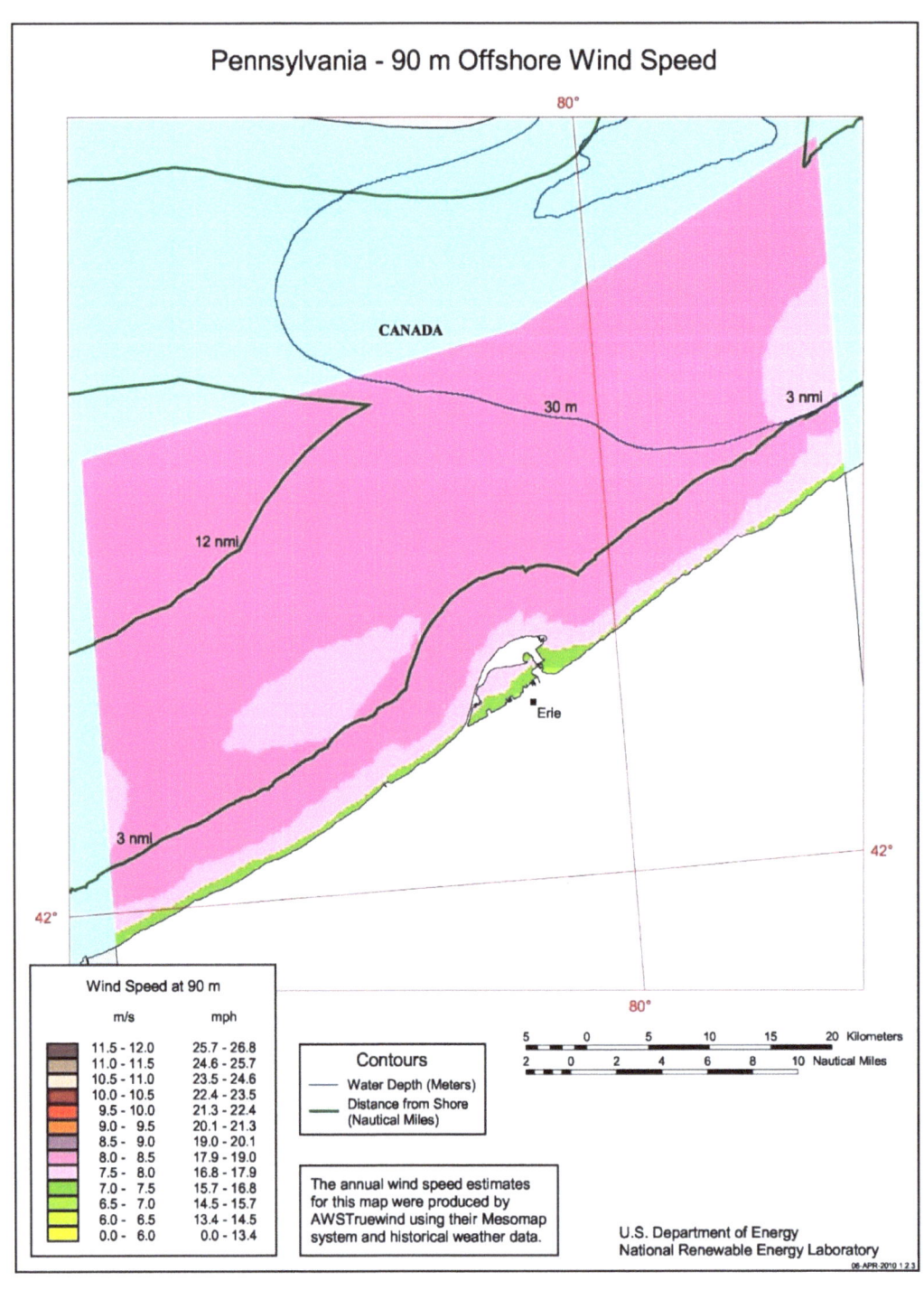

Figure B21. Pennsylvania detailed map

Table B22. Rhode Island offshore wind resource by wind speed interval, water depth and distance from shore within 50 nm of shore.

	Distance from Shore (nm)								
	0 - 3			3 - 12			12 - 50		
Depth Category	Shallow (0 - 30 m)	Transitional (30 - 60m)	Deep (> 60m)	Shallow (0 - 30 m)	Transitional (30 - 60m)	Deep (> 60m)	Shallow (0 - 30 m)	Transitional (30 - 60m)	Deep (> 60m)
90 m Wind Speed Interval (m/s)	Area km² (MW)	Area km² (MW)	Area km² (MW)	Area km² (MW)	Area km² (MW)	Area km² (MW)	Area km² (MW)	Area km² (MW)	Area km² (MW)
7.0 - 7.5	216 (1,082)	8 (40)	0 (0)	0 (0)	0 (0)	0 (0)	0 (0)	0 (0)	0 (0)
7.5 - 8.0	123 (616)	3 (14)	0 (0)	0 (0)	0 (0)	0 (0)	0 (0)	0 (0)	0 (0)
8.0 - 8.5	140 (700)	40 (199)	0 (0)	35 (173)	69 (346)	0 (0)	0 (0)	0 (0)	0 (0)
8.5 - 9.0	120 (602)	93 (465)	0 (0)	184 (918)	274 (1,371)	0 (0)	0 (0)	0 (0)	0 (0)
9.0 - 9.5	54 (269)	18 (88)	0 (0)	176 (880)	783 (3,914)	0 (0)	0 (0)	430 (2,151)	1 (6)
9.5 - 10.0	0 (0)	0 (0)	0 (0)	0 (0)	6 (29)	0 (0)	0 (0)	967 (4,836)	1,387 (6,934)
>10.0	0 (0)	0 (0)	0 (0)	0 (0)	0 (0)	0 (0)	0 (0)	0 (0)	0 (0)
Total >7.0	653 (3,267)	161 (806)	0 (0)	394 (1,970)	1,132 (5,660)	0 (0)	0 (0)	1,397 (6,987)	1,388 (6,940)

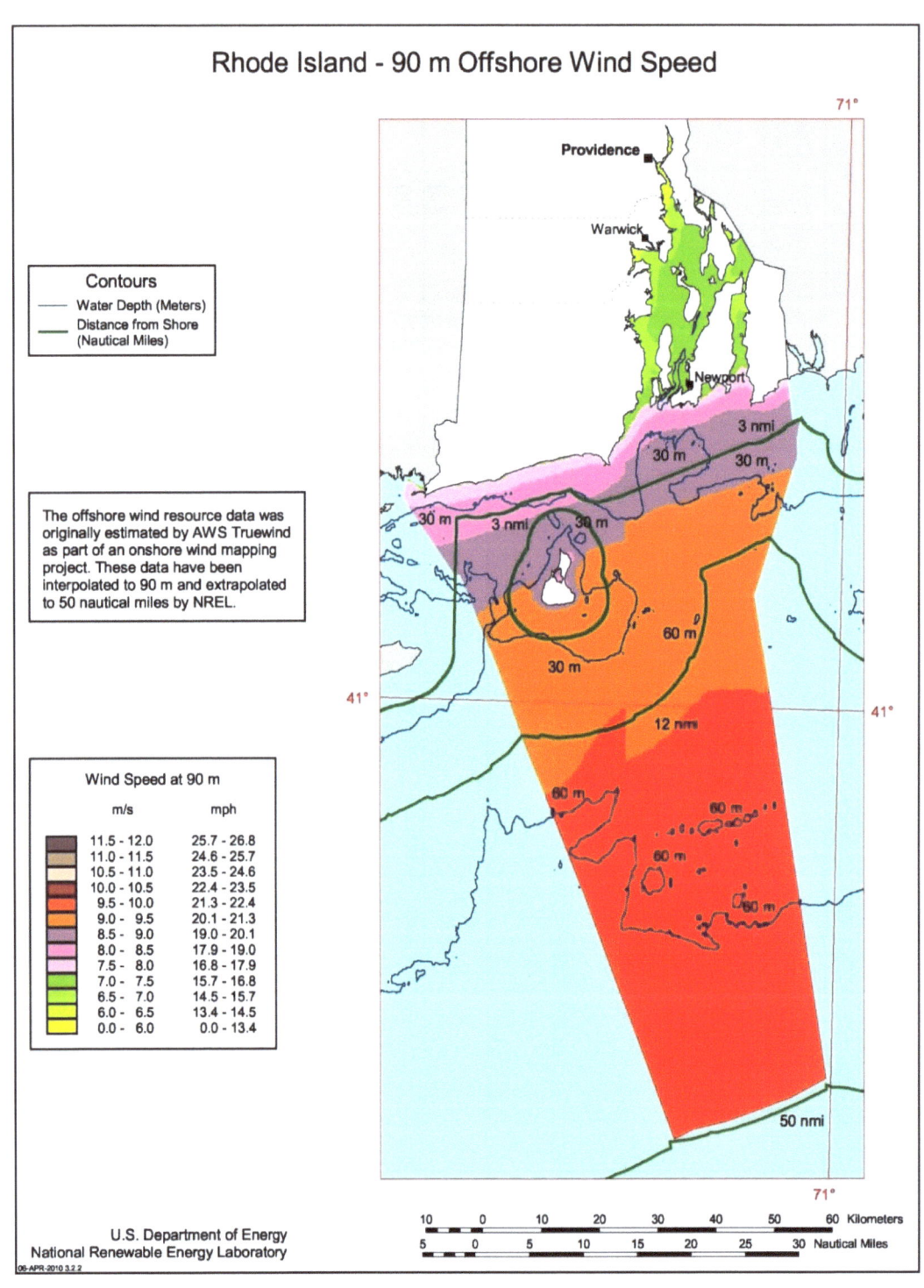

Figure B22. Rhode Island detailed map

Table B23. South Carolina offshore wind resource by wind speed interval, water depth and distance from shore within 50 nm of shore.

	Distance from Shore (nm)								
	0 - 3			3 - 12			12 - 50		
Depth Category	Shallow (0 - 30 m)	Transitional (30 - 60m)	Deep (> 60m)	Shallow (0 - 30 m)	Transitional (30 - 60m)	Deep (> 60m)	Shallow (0 - 30 m)	Transitional (30 - 60m)	Deep (> 60m)
90 m Wind Speed Interval (m/s)	Area km^2 (MW)	Area km^2 (MW)	Area km^2 (MW)	Area km^2 (MW)	Area km^2 (MW)	Area km^2 (MW)	Area km^2 (MW)	Area km^2 (MW)	Area km^2 (MW)
7.0 - 7.5	848 (4,241)	0 (0)	0 (0)	608 (3,042)	0 (0)	0 (0)	0 (0)	0 (0)	0 (0)
7.5 - 8.0	594 (2,968)	0 (0)	0 (0)	3,054 (15,269)	0 (0)	0 (0)	4,268 (21,338)	287 (1,435)	0 (0)
8.0 - 8.5	23 (115)	0 (0)	0 (0)	1,609 (8,047)	0 (0)	0 (0)	4,151 (20,757)	3,926 (19,628)	674 (3,372)
8.5 - 9.0	0 (0)	0 (0)	0 (0)	0 (0)	0 (0)	0 (0)	2,027 (10,135)	3,110 (15,548)	870 (4,349)
9.0 - 9.5	0 (0)	0 (0)	0 (0)	0 (0)	0 (0)	0 (0)	0 (0)	0 (0)	0 (0)
9.5 - 10.0	0 (0)	0 (0)	0 (0)	0 (0)	0 (0)	0 (0)	0 (0)	0 (0)	0 (0)
>10.0	0 (0)	0 (0)	0 (0)	0 (0)	0 (0)	0 (0)	0 (0)	0 (0)	0 (0)
Total >7.0	1,465 (7,323)	0 (0)	0 (0)	5,272 (26,358)	0 (0)	0 (0)	10,446 (52,230)	7,322 (36,611)	1,544 (7,722)

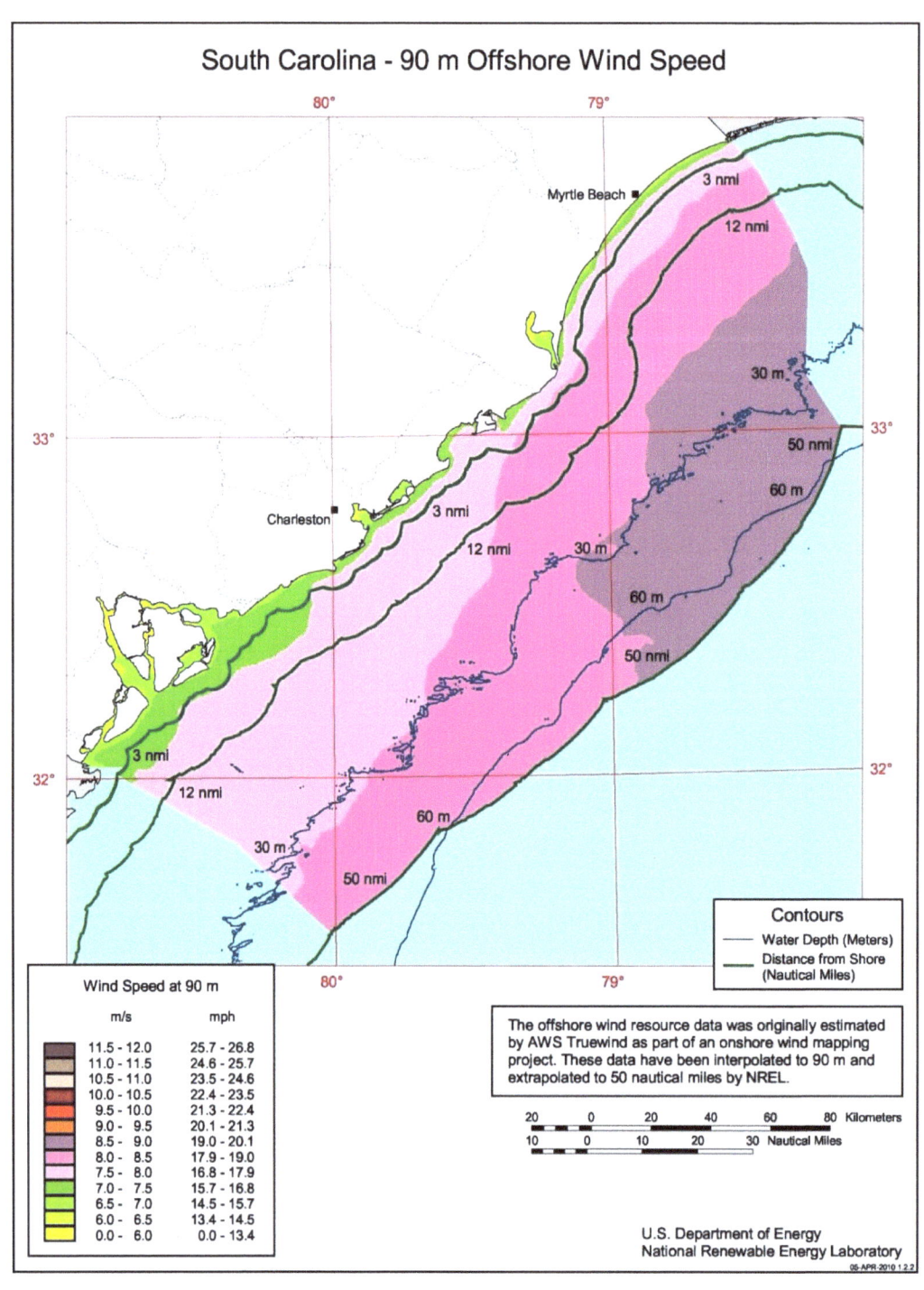

Figure B23. South Carolina detailed map

Table B24. Texas offshore wind resource by wind speed interval, water depth and distance from shore within 50 nm of shore.

	Distance from Shore (nm)								
	0 – 9*			9 - 12			12 - 50		
Depth Category	Shallow (0 - 30 m)	Transitional (30 - 60m)	Deep (> 60m)	Shallow (0 - 30 m)	Transitional (30 - 60m)	Deep (> 60m)	Shallow (0 - 30 m)	Transitional (30 - 60m)	Deep (> 60m)
90 m Wind Speed Interval (m/s)	Area km² (MW)	Area km² (MW)	Area km² (MW)	Area km² (MW)	Area km² (MW)	Area km² (MW)	Area km² (MW)	Area km² (MW)	Area km² (MW)
7.0 - 7.5	1,786 (8,928)	0 (0)	0 (1)	96 (481)	0 (0)	0 (0)	137 (686)	0 (0)	0 (0)
7.5 - 8.0	9,046 (45,230)	0 (0)	0 (0)	1,165 (5,826)	0 (0)	0 (0)	8,045 (40,226)	6,175 (30,873)	392 (1,958)
8.0 - 8.5	5,929 (29,643)	125 (627)	0 (0)	742 (3,711)	137 (684)	0 (0)	584 (2,918)	4,443 (22,214)	4,597 (22,985)
8.5 - 9.0	3,798 (18,990)	601 (3,005)	0 (0)	43 (215)	945 (4,726)	0 (0)	0 (0)	3,211 (16,055)	3,675 (18,373)
9.0 - 9.5	0 (0)	0 (0)	0 (0)	0 (0)	0 (0)	0 (0)	0 (0)	0 (0)	0 (0)
9.5 - 10.0	0 (0)	0 (0)	0 (0)	0 (0)	0 (0)	0 (0)	0 (0)	0 (0)	0 (0)
>10.0	0 (0)	0 (0)	0 (0)	0 (0)	0 (0)	0 (0)	0 (0)	0 (0)	0 (0)
Total >7.0	20,558 (102,791)	726 (3,632)	0 (1)	2,046 (10,232)	1,082 (5,410)	0 (0)	8,766 (43,829)	13,828 (69,142)	8,663 (43,316)

*Federal waters begin at 9 nm for Texas

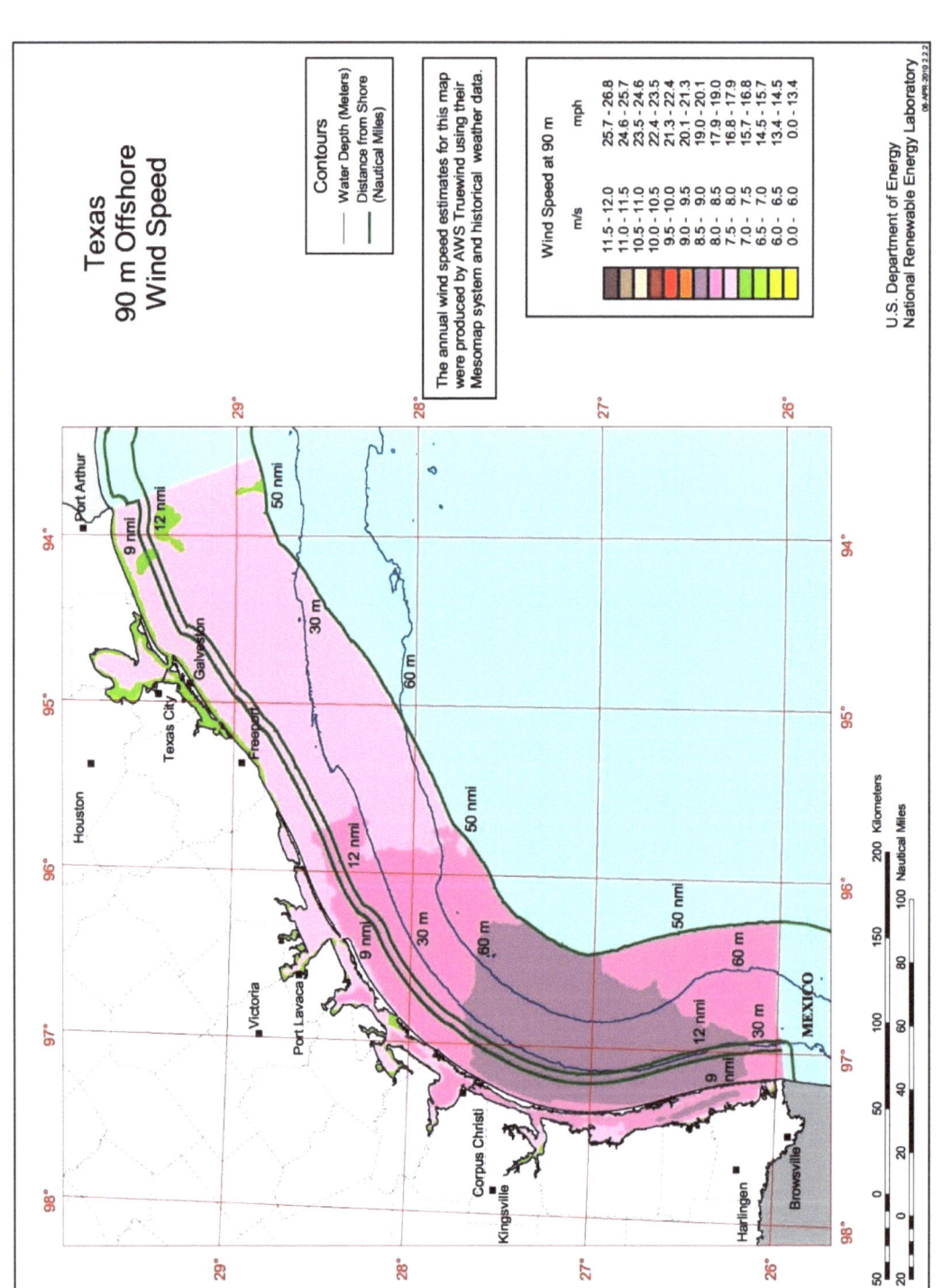

Figure B24. Texas detailed map

90

Table B25. Virginia offshore wind resource by wind speed interval, water depth and distance from shore within 50 nm of shore.

	Distance from Shore (nm)								
	0 - 3			3 - 12			12 - 50		
Depth Category	Shallow (0 - 30 m)	Transitional (30 - 60m)	Deep (> 60m)	Shallow (0 - 30 m)	Transitional (30 - 60m)	Deep (> 60m)	Shallow (0 - 30 m)	Transitional (30 - 60m)	Deep (> 60m)
90 m Wind Speed Interval (m/s)	Area km² (MW)	Area km² (MW)	Area km² (MW)	Area km² (MW)	Area km² (MW)	Area km² (MW)	Area km² (MW)	Area km² (MW)	Area km² (MW)
7.0 - 7.5	889 (4,446)	0 (0)	0 (0)	0 (0)	0 (0)	0 (0)	0 (0)	0 (0)	0 (0)
7.5 - 8.0	3,606 (18,028)	15 (77)	0 (0)	37 (185)	0 (0)	0 (0)	0 (0)	0 (0)	0 (0)
8.0 - 8.5	1,136 (5,679)	2 (10)	0 (0)	2,986 (14,929)	0 (0)	0 (0)	2,356 (11,780)	69 (345)	0 (0)
8.5 - 9.0	0 (0)	0 (0)	0 (0)	24 (119)	0 (0)	0 (0)	2,030 (10,152)	4,840 (24,199)	900 (4,500)
9.0 - 9.5	0 (0)	0 (0)	0 (0)	0 (0)	0 (0)	0 (0)	0 (0)	0 (0)	0 (0)
9.5 - 10.0	0 (0)	0 (0)	0 (0)	0 (0)	0 (0)	0 (0)	0 (0)	0 (0)	0 (0)
>10.0	0 (0)	0 (0)	0 (0)	0 (0)	0 (0)	0 (0)	0 (0)	0 (0)	0 (0)
Total >7.0	5,630 (28,152)	17 (87)	0 (0)	3,047 (15,233)	0 (0)	0 (0)	4,386 (21,932)	4,909 (24,544)	900 (4,500)

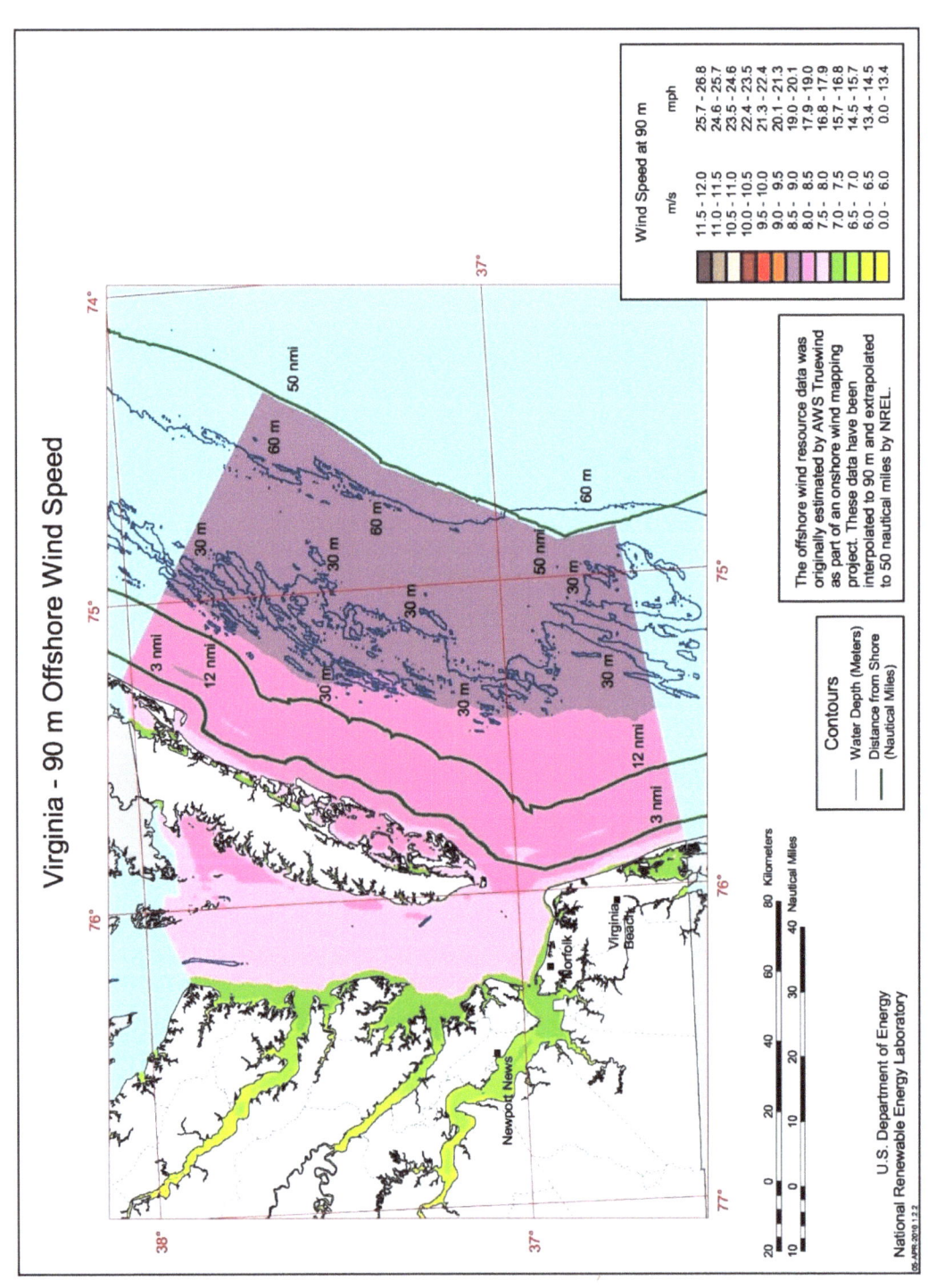

Figure B25. Virginia detailed map

Table B26. Washington offshore wind resource by wind speed interval, water depth and distance from shore within 50 nm of shore.

Depth Category / 90 m Wind Speed Interval (m/s)	0 - 3			3 - 12			12 - 50		
	Shallow (0 - 30 m) Area km² (MW)	Transitional (30 - 60 m) Area km² (MW)	Deep (> 60m) Area km² (MW)	Shallow (0 - 30 m) Area km² (MW)	Transitional (30 - 60m) Area km² (MW)	Deep (> 60m) Area km² (MW)	Shallow (0 - 30 m) Area km² (MW)	Transitional (30 - 60m) Area km² (MW)	Deep (> 60m) Area km² (MW)
7.0 - 7.5	623 (3,115)	133 (667)	57 (286)	232 (1,159)	350 (1,751)	177 (884)	0 (0)	0 (0)	0 (1)
7.5 - 8.0	371 (1,853)	0 (0)	0 (0)	203 (1,015)	842 (4,208)	1,173 (5,865)	0 (0)	0 (0)	2,033 (10,166)
8.0 - 8.5	62 (308)	0 (0)	0 (0)	212 (1,059)	978 (4,889)	475 (2,375)	0 (0)	20 (98)	16,515 (82,574)
8.5 - 9.0	0 (0)	0 (0)	0 (0)	0 (0)	0 (0)	0 (0)	0 (0)	0 (0)	0 (0)
9.0 - 9.5	0 (0)	0 (0)	0 (0)	0 (0)	0 (0)	0 (0)	0 (0)	0 (0)	0 (0)
9.5 - 10.0	0 (0)	0 (0)	0 (0)	0 (0)	0 (0)	0 (0)	0 (0)	0 (0)	0 (0)
>10.0	0 (0)	0 (0)	0 (0)	0 (0)	0 (0)	0 (0)	0 (0)	0 (0)	0 (0)
Total >7.0	1,055 (5,276)	133 (667)	57 (286)	647 (3,233)	2,170 (10,849)	1,825 (9,124)	0 (0)	20 (98)	18,548 (92,741)

Distance from Shore (nm)

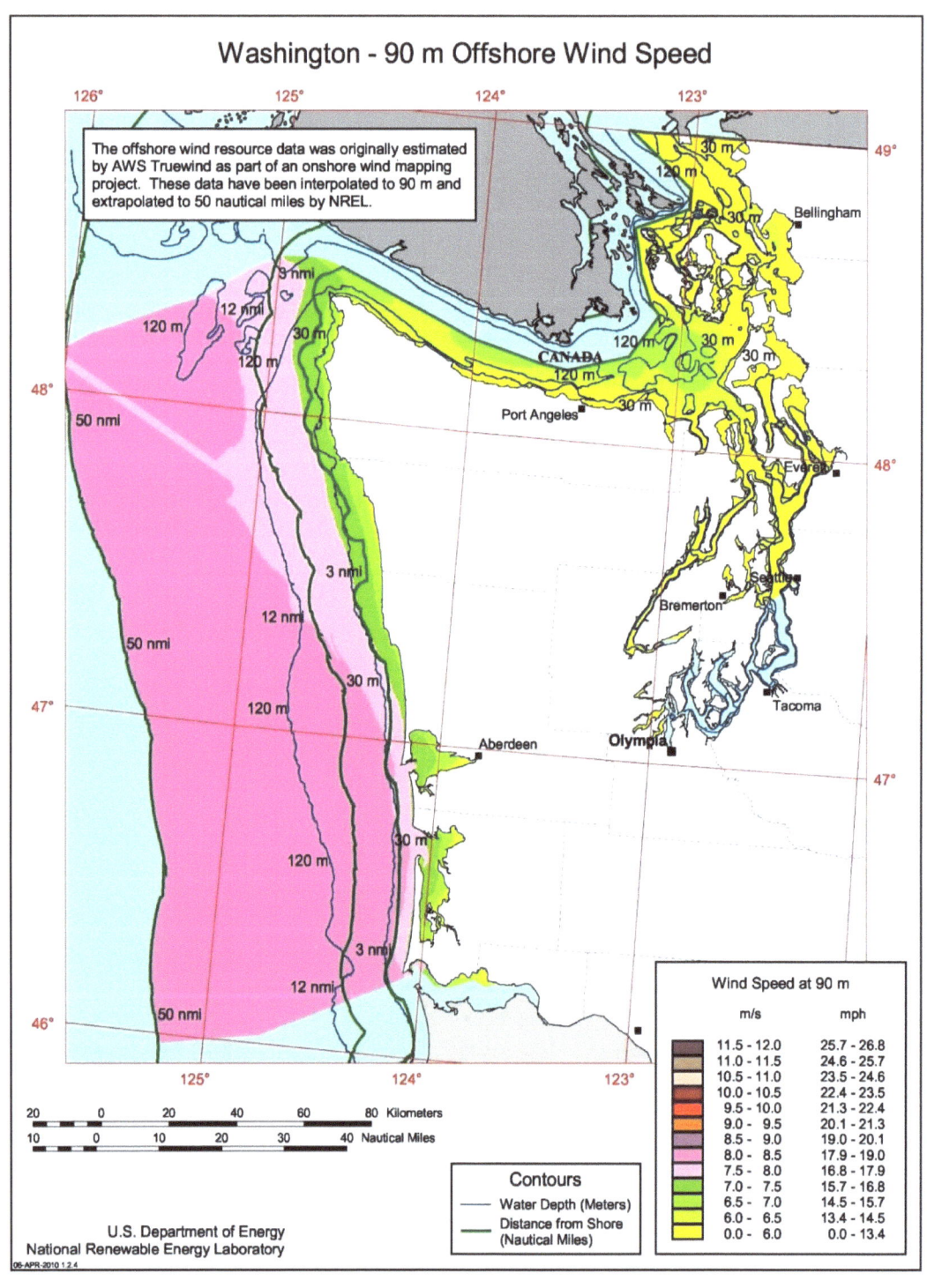

Figure B26. Washington detailed map

Table B27. Wisconsin offshore wind resource by wind speed interval, water depth and distance from shore within 50 nm of shore.

	Distance from Shore (nm)								
	0 - 3			3 - 12			12 - 50		
Depth Category	Shallow (0 - 30 m)	Transitional (30 - 60m)	Deep (> 60m)	Shallow (0 - 30 m)	Transitional (30 - 60m)	Deep (> 60m)	Shallow (0 - 30 m)	Transitional (30 - 60m)	Deep (> 60m)
90 m Wind Speed Interval (m/s)	Area km² (MW)	Area km² (MW)	Area km² (MW)	Area km² (MW)	Area km² (MW)	Area km² (MW)	Area km² (MW)	Area km² (MW)	Area km² (MW)
7.0 - 7.5	1,054 (5,269)	268 (1,341)	146 (729)	350 (1,749)	617 (3,083)	1,085 (5,425)	0 (0)	34 (171)	162 (810)
7.5 - 8.0	1,417 (7,086)	235 (1,177)	24 (119)	362 (1,808)	223 (1,115)	779 (3,896)	0 (0)	0 (0)	364 (1,821)
8.0 - 8.5	831 (4,153)	373 (1,863)	0 (0)	451 (2,257)	1,386 (6,930)	2,539 (12,695)	0 (0)	21 (107)	2,160 (10,800)
8.5 - 9.0	26 (128)	8 (41)	0 (0)	12 (58)	301 (1,504)	1,347 (6,737)	0 (0)	63 (314)	6,661 (33,305)
9.0 - 9.5	0 (0)	0 (0)	0 (0)	0 (0)	0 (0)	0 (0)	0 (0)	0 (0)	0 (0)
9.5 - 10.0	0 (0)	0 (0)	0 (0)	0 (0)	0 (0)	0 (0)	0 (0)	0 (0)	0 (0)
>10.0	0 (0)	0 (0)	0 (0)	0 (0)	0 (0)	0 (0)	0 (0)	0 (0)	0 (0)
Total >7.0	3,327 (16,636)	884 (4,422)	169 (847)	1,175 (5,873)	2,526 (12,632)	5,751 (28,753)	0 (0)	118 (592)	9,347 (46,737)

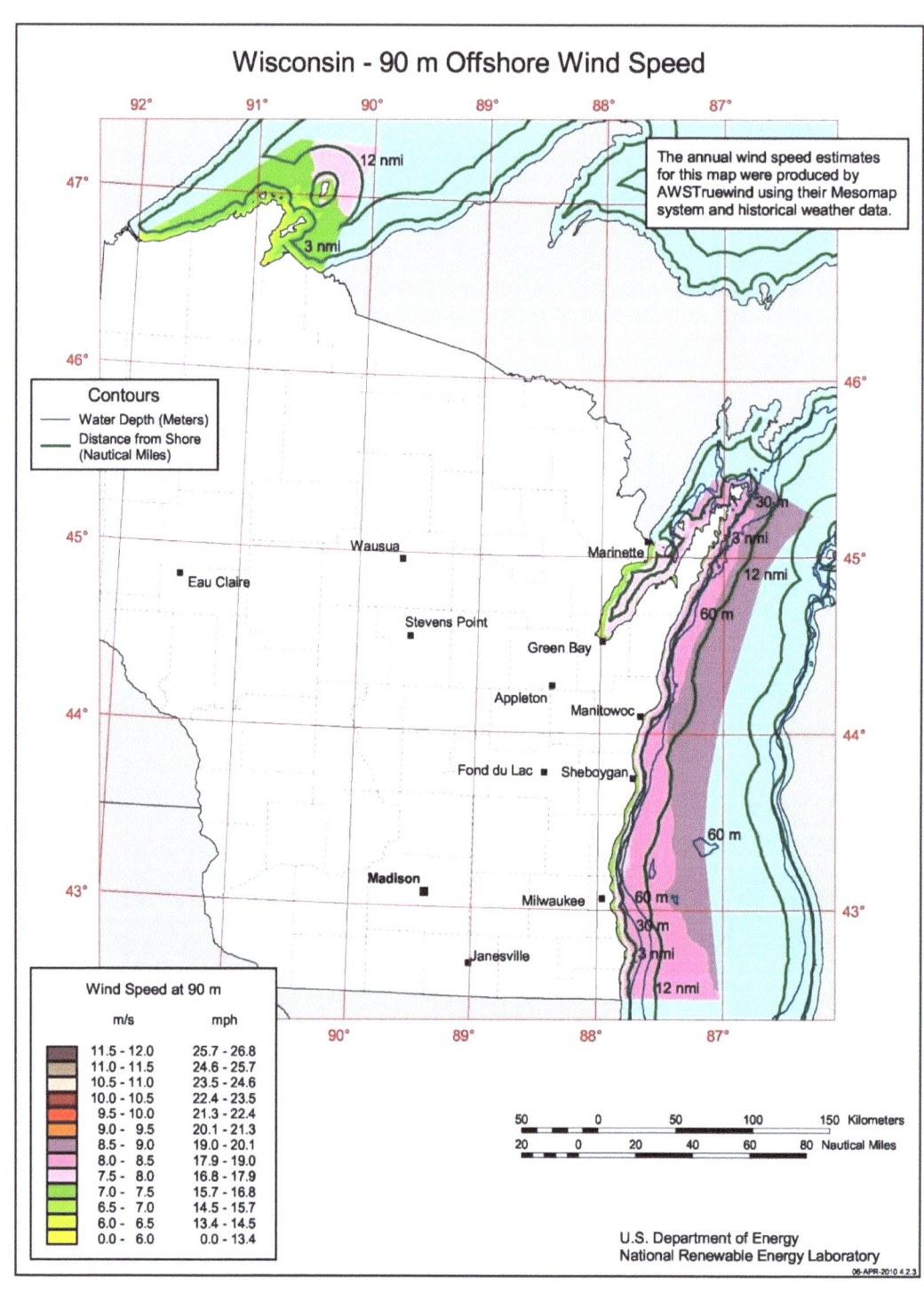

Figure B27. Wisconsin detailed map

www.ingramcontent.com/pod-product-compliance
Lightning Source LLC
Chambersburg PA
CBHW050728180526
45159CB00003B/1164